Understanding
Boat
Communications

OTHER BOOKS BY JOHN C. PAYNE

The Fisherman's Electrical Manual

The Great Cruising Cookbook: an International Galley Guide

Marine Electrical and Electronics Bible, Second Edition

Motorboat Electrical and Electronics Manual

Understanding Boat Batteries and Battery Charging

Understanding Boat Corrosion, Lightning Protection and Interference

Understanding Boat Diesel Engines

Understanding Boat Electronics

Understanding Boat Wiring

Visit John C. Payne's Website at www.fishingandboats.com

Enroll in John C. Payne's Marine Electrical School at
www.fishingandboats.com

Understanding Boat Communications

JOHN C. PAYNE

SHERIDAN HOUSE

This edition first published 2006 by
Sheridan House Inc.
145 Palisade Street
Dobbs Ferry, NY 10522
www.sheridanhouse.com

Library of Congress Cataloging-in-Publication Data
Payne, John C.
    Understanding boat communications / John C. Payne.
        p. cm.
    Includes index.
    ISBN 1-57409-229-4 (pbk. : alk. paper)
    1. Electronics in navigation. 2. Mobile communication systems.
    3. Wireless communication systems. I. Title.

    VK397.P39 2006
    623.8'56—dc22   2006002189

ISBN 1-57409-229-4

Printed in the United States of America

CONTENTS

1. GMDSS COMMUNICATIONS

What is the Global Maritime Distress and Safety System (GMDSS)?

The primary function of GMDSS is to coordinate and facilitate Search and Rescue (SAR) operations, by both shore authorities and vessels, with the shortest possible delay and maximum efficiency. It also provides efficient urgency and safety communications, and broadcast of Maritime Safety Information (MSI) such as navigational and meteorological warnings, forecasts, and other urgent safety information. MSI is transmitted via NAVTEX, International SafetyNet on INMARSAT C, and some radio telex services. GMDSS was fully implemented in 1999 for all commercial vessels exceeding 300 GRT.

What are the GMDSS operational details?

This is a system with worldwide communications coverage. It is achieved by using a combination of INMARSAT satellite communications and terrestrial systems. All of these systems have range limitations that have resulted in the designation of four sea areas, which defines communications system requirements.

What is Area A1?

Area 1 is within shore-based VHF radio range. Distance is in the range of 20-100 nm. Radio required is VHF operating on Channel 70 for DSC, and Channel 16 radiotelephone. EPIRB required is 406 MHz or L-band unit (1.6 GHz). A VHF EPIRB is required. Survival craft require a 9-GHz radar transponder and portable VHF radio (with Channel 16 and one other frequency).

What is Area A2?

Area 2 is within shore-based MF radio range. Distance is in the range of 100-300 nm. Radios required are MF (2187.5 kHz DSC) and 2812 kHz radiotelephone, 2174.5 NBDP, and NAV-TEX on 518 kHz. Also needed are the same VHF requirements as Area A1. EPIRB required is 406 MHz or L-band (1.6 GHz). Survival craft requirements are the same as in Area A1.

What is Area A3?

Area 3 is within geostationary satellite range (INMARSAT). Distance is in the range of 70°N-70°S. Radios required are MF and VHF as above and satellite (with 1.5-1.6 GHz alerting), or as per Areas A1 and A2 plus HF (all frequencies). Survival craft requirements are the same as in Area A1.

What is Area A4?

Area 4 includes the other areas (beyond INMARSAT range). Distance north of 70°N and south of 70°S. Radios required are HF, MF, and VHF. EPIRB required is 406 MHz. Survival craft requirements are the same as in Area A1.

What are the GMDSS radio distress communications frequencies?

The frequencies designated for use under GMDSS are as follows:

1. VHF   DSC Channel 70, Channel 16, Channel 06 intership, Channel 13 intership MSI.

2. MF   DSC 2187.5 kHz, and 2182 kHz.

3. HF4   DSC 4207.5 kHz, and 4125 kHz.

4. HF6   DSC 6312 kHz, and 6215 kHz (CH421).

5. HF8   DSC 8414.5 kHz, and 8291 kHz (CH833).

6. HF12   DSC 12577 kHz, and 12290 kHz (CH1221).

7. HF16   DSC 16804.5 kHz, and 16420 kHz.

What is digital selective calling (DSC)?

DSC is a primary component of GMDSS and is used to transmit distress alerts and appropriate acknowledgments. DSC was introduced to improve accuracy, transmission, and reception of distress calls. VHF Channel 70 is the nominated DSC channel.

1. DSC has the advantage that digital signals in radio communications are at least 25% more efficient than voice transmissions, as well as significantly faster. A DSC VHF transmission typically takes around a second, and MF/HF takes approximately 7 seconds, both depending on the DSC call type.

DSC Radio
Courtesy Standard Horizon

2. DSC requires the use of encoders/decoders, or additional add-on modules to existing equipment. A dedicated DSC watch receiver is required to continuously monitor the specified DSC distress frequency. Affordable VHF DSC radio equipment is a priority for small vessels and Class D controllers are now available.

3. DSC equipment enables the transmission of digital information based on four priority groupings, Distress, Urgency, Safety, and Routine. The information can be selectively addressed to all stations, to a specific station, or to a group of stations. To perform this selective transmission and reception of messages, every station must possess what is called a Maritime Mobile Selective-call Identity Code (MMSI). Note that distress "Mayday" messages are automatically dispatched to all stations. A DSC distress alert message is configured to contain the transmitting vessel identity (the MMSI nine-digit code number), the time, the nature of the distress, and the vessel position where interfaced with a GPS. After transmission of a distress alert, it is repeated a few seconds later to ensure that the transmission is successful.

What is the GMDSS distress call (alert) sequence?

1. **Distress Alert.** This is usually activated from a vessel to shore. For motorboats this is usually via terrestrial radio, and larger vessels use satellites. Ships in the area may hear an alert, although a shore-based Rescue Coordination Center (RCC) will be responsible for responding to and acknowledging receipt of the alert. Alerts may be activated via an INMARSAT A, B, or C terminal, via COSPAS/SARSAT EPIRB (243/406 MHz), or via an INMARSAT E EPIRB. DSC VHF or MF/HF can also activate alerts.

2. **Distress Relay.** On receipt and acknowledgment of alert, the RCC will relay the alert to vessels in the geographical area concerned. This targets the resources available and does not involve vessels outside the distress vessel area. Vessels in the area of distress can receive appropriate alerts via INMARSAT A, B, or C terminals, DSC VHF or MF/HF radio equipment, or via NAVTEX MSI. On reception of a distress relay the vessels concerned must contact the RCC to offer assistance.

3. **Search and Rescue.** In the SAR phase of the rescue, the previous one-way communications switch over to two-way for effective coordination of both aircraft and vessels. The frequencies used are as outlined previously.

4. **Rescue Scene Communications.** Local communications are maintained using short-range terrestrial MF or VHF on the specified frequencies. Local communications take place using either satellite or terrestrial radio links.

5. **Distress Vessel Location.** A Search and Rescue Transponder (SART), and/or the 121.5MHz homing frequency of an EPIRB assist in determining the precise location of the vessel in distress.

About false alerts and system coverage

GMDSS is relatively new, and currently the false alert rate is around 85%. False alerts are placing a load on SAR services. They are generally caused by operator errors and incorrect equipment operation. Another cause of false alerts is the improper acknowledgment of distress alerts leading to excessive DSC calls. Training and experience of equipment operation is essential to reduce these problems.

About GMDSS and pleasure boats

The installation of GMDSS is not compulsory for pleasure boats, but due to its universal implementation on commercial vessels, boats will be forced to install partial GMDSS equipment simply to remain "plugged in" to the system. GMDSS will certainly maximize SAR situations for boats so in most cases it will enhance offshore safety. GMDSS equipment will accurately identify your own boat, current position, and type of emergency, and this information will be broadcast automatically. What you get is automatic activation of alarms at coast stations and on other vessels simply by pushing one button. Just as GPS, electronic charting, and the EPIRB have opened up the world to cruisers, so will GMDSS improve significantly sea safety. As a minimum the following equipment is required for an offshore trip. Few will be able to invest in full INMARSAT terminals. A more advanced training course and operation certificate is also required and is being run in many locations.

1. NAVTEX receiver

2. 406 EPIRB (correctly registered)

3. VHF DSC (Class D controller)

4. VHF (approved handheld type)

5. SART (optional but desirable)

6. MF/HF Class E DSC (optional but desirable)

2. EPIRBs, PLBs, SARTs, RTEs

All About EPIRBs

GMDSS incorporates the COSPAS/SARSAT system as an integral part of the distress communications system. The acronym is based on the former Soviet "Space System for Search of Distress Vessels" and the American "Search and Rescue Satellite Aided Tracking." Under GMDSS if a vessel does not carry a satellite L-band EPIRB in sea areas A1, A2, and A3 (described earlier), a 406 MHz EPIRB is required. This unit must have hydrostatic release and float-free capability. The system is a worldwide satellite-assisted SAR system for location of distress transmissions emitted by EPIRBs on the 121.5/243 MHz and 406 MHz frequencies. 121.5 kHz is an aircraft homing frequency and 243 MHz is a military distress frequency that enables military aircraft to assist in SAR operations. The Emergency Position Indicating Radio Beacon (EPIRB) is an essential item of safety equipment for any offshore vessel.

About EPIRB detection

Earlier EPIRB units relied solely on over-flying aircraft for detection of signals and relay of the position to appropriate SAR authorities; the new systems utilize satellites. Orbit time is 100 minutes using four COSPAS satellites so a delay in transmission up to 4 hours near the equator can occur. Accuracy is 2km and uses Doppler effect; it usually requires 2 satellite passes. The system uses a store and forward system so the satellite stores and downloads distress data when in view of a Local User Terminal (LUT).

How accurate are EPIRBs?

Accuracy of the system improved from approximately 10 nm for 121.5/243 MHz units to 3 nm for a 406 MHz unit. Note that the 406 MHz units are far more effective at lower latitudes than the 121.5/243 MHz units, and the latter are being phased out. These were not useful in mid ocean and had a high false alarm rate so it was usual to wait for two separate satellite hits before activating SAR. There is no vessel specific identification, and error is around 20km. Aviation frequency is 121.5 kHz, and military 243 MHz. 121.5 kHz is now primarily used on personal locator beacons only.

What are Class A and Class B units?

The Class A EPIRB transmits on 121.5 and 243.0 MHz, floats free and turns on automatically. A Class B unit operates on 121.5 and 243.0 MHz, is manually deployed and turned on. The NOAA will stop monitoring 121.5 MHz signals starting in 2009.

How does an EPIRB float free?

A mechanism called a Hydrostatic Release Unit (HRU) is used. One type of release mechanism incorporates a small pyrotechnic charge. When the depth reaches 1 to 4 meters a pressure diaphragm activates the charge which then drives the cutting blade through the plastic retaining bolt securing the EPIRB. HRU must comply with USCG and SOLAS requirements.

What about strobe lights?

Some EPIRBs incorporate a high light intensity xenon strobe light. Most of these have a flash frequency of 23 flashes per minute.

What are satellite (L-Band) EPIRBs?

This system, developed by the European Space Agency, will alert rescue services in distress within 2 minutes, rather than in hours as with current systems. The new system combines position determination along with a distress signal using the INMARSAT geostationary satellites. The system uses special EPIRBs that incorporate GPS receivers and ensure a position fix within 200 meters. The distress signal transmits via one of four Land Earth Stations (LES) and landline links with appropriate rescue coordination centers. Recent testing shows an average 5-minute delay from activation to reception by rescue services.

About 406 MHz EPIRBs

The 406 MHz units have a unique identification code, and information is usually programmed at time of sale, with MMSI or registered serial numbers. Some units also have integral strobes and all incorporate 121.5 MHz for aircraft homing signal purposes. Accuracy is typically around 2-3 nm.

1. **Category 1.** These are float-free units and switch on automatically.

2. **Category 2.** These are manual bracket units that are manually deployed and automatically switch on.

What is a GPS EPIRB?

Some EPIRBs also incorporate an integral 12 channel GOS receiver. The receiver gives precise latitude and longitude that is transmitted, down to 100-foot accuracy, which can cut down the localization part of any search and rescue operation.

About 406 MHz EPIRB registration

If you buy an EPIRB or acquire a vessel with a 406 MHz EPIRB, you must register the EPIRB unit properly and provide all of the appropriate data, including its unique identification number (or MMSI). This is mandatory. Registration should be done immediately upon purchase. Failure to do this can cause havoc if you use it, because a vessel may be incorrectly identified or, worse still, not identified at all, which could seriously jeopardize your rescue. Bad information means very bad rescue problems for everyone. The process requires vessel details along with contact details that are used to authenticate distress transmissions.

If you have not registered, contact the organizations listed:

1. **United States of America.** NOAA/NESDIS (tel: 1-301-457 5428). Additional information on registration (tel: 1-302-763 4680 and 888-212-SAVE or www.sarsat.noaa.gov.) Also visit www.beaconregistration.noaa.gov for online registration.

2. **United Kingdom.** EPIRB Registry, Marine Safety Agency (tel: 44-1326 211569 or www.msa.co.uk).

3. **Canada.** Canadian EPIRB Registry Director, Search and Rescue, Canadian Coastguard, (tel: 1-613-990 3124).

4. **Australia.** Maritime Rescue Co-ordination Center, Australian Maritime Safety Authority. (tel: 61-2-6230 6811.) rrcaus@amsa.gov.au and www.amsa.com.au.

5. **New Zealand.** CAA. (tel: 64-4-5600400.)

About the EPIRB activation sequence

When an EPIRB is activated the following sequence of events occurs:

1. A satellite detects the distress transmission. With 243/ 121.5 MHz units a satellite and the EPIRB must be simultaneously in view of the Local User Terminal (LUT).

2. The detected signal is then downloaded to a LUT. (In 406 MHz units the satellite stores the message and downloads to the next LUT in view).

3. The LUT automatically computes the position of the distress transmission. The distress information is then passed to a Mission Control Center (MCC) before going to a Rescue Control Center (RCC) and then to SAR aircraft and vessels.

About EPIRB operation

Do not operate an EPIRB except in a real emergency, because you could initiate a rescue operation. Do not even operate it for just a short period of time and then switch it off, because authorities may assume your vessel went down quickly before circumstances stopped transmission. False alarms are seen as wasting taxpayers' money and reflect very badly on boaters as a whole. If you activate your EPIRB during an emergency, once rescued, do not leave the EPIRB in the raft or floating as the beacon may continue to transmit for some time.

How long will the rescue take?

There is a mistaken belief that rescues are instantaneous after activation of an EPIRB. In reality, there is a notification time lag that can average up to 6 hours or more from the detection of a signal and physical location for a 121.5/243 MHz unit, although the position within 12 nm is usually confirmed in less than 2 hours. This is dependent on suitable aircraft, weather conditions, and SAR coordinator response times. For a 406 MHz EPIRB it is around 40 minutes, within 2 nm and GPS equipped units this comes down to 4 minutes within 100 meters. Every LUT has a "footprint" coverage area, and the closer you are to the edge of that footprint, the longer the delay. Time lags depend on intervals between satellite passes over a given location. There are six polar orbiting satellites and, although random in orbit, their tracks are predictable. If you have to activate, be patient and wait. Remember, you are not a survivor until you're on the deck of a rescue vessel or in the helicopter. Priority one is a survival training course. Have you done one? Have you evaluated and planned a helicopter evacuation procedure?

About battery life and transmit times

Much concern has been raised over battery transmit life after activation. Always ensure that the battery pack is replaced within the listed expiration date. Nominally a lithium battery has a life of 4 to 6 years depending on the manufacturer, some quote a storage life of 10-12 years. Typical transmit times are 24-48 hours at 5 watt power output which is the normal rating.

What EPIRB maintenance is required

The only maintenance required is to test the EPIRB using the self-test function every six months in accordance with the manufacturer's instructions. Do not self-test by activating the EPIRB distress function. Do not drop the unit unless it is in the water or damage may occur.

Personal locator beacons (PLBs)

These are not GMDSS equipment. The PLB is essentially a miniature EPIRB. Many operate on 121.5 MHz, which is the frequency used for homing in by SAR vessels and aircraft, but 406 MHz units are now available also. Due to their small size they can be attached to wet weather gear, or carried in a pocket or grab bag. The PLB is not as accurate as other units and will localize your position to around 12 nm, because the transmitters are line of sight only. The 406 MHz PLB will give accuracy to 3 nm. There are also small handheld PLB units that are 406 MHz and GPS equipped. Some units are configured to activate in water, and most operate for at least a 24-hour period and some work up to 48 hours. The PLB is not a substitute for a vessel specific 243/121.5 MHz or 406 MHz EPIRB.

Personal locator beacon direction finders

As PLBs are known as Crew Overboard Beacons some boats also have a portable direction finder. This is a small handheld device that allows the boat to find the 121.5 MHz homing signal emitted from the PLB.

Radar target enhancers (RTEs)

These are not GMDSS equipment. The operation of these devices includes the reception of an incoming radar signal, the amplification of that pulse, and the retransmission of the pulse back to the radar signal source. This has to occur simultaneously and at the same frequency. The returned signal is displayed in enhanced form, with the relatively small return of the boat appearing significantly larger than it actually is. As consistency of the radar signal is a major factor with vessel radars and with most commercial vessels having ARPA (Automatic Radar Plotting Aid) radars, this allows them to see you and activate an alarm if a collision risk exists.

How effective is an RTE?

Some new units claim a target enhancement factor of 6 to 8 times greater than the actual reflected image. This obviously has the advantage of displaying strong and consistent echoes on radar screens. Effectiveness depends on the incoming radar signal strength, height at which the RTE is installed, and the height of the other vessel's radar above sea level. The Ocean Sentry unit operates either in standby or transponder modes. In standby mode, the unit is activated only when a radar signal is present. These units operate in response to 3-cm X-band radars only, not S-band. The effective range is typically around 12 nm, but not less than around 3 nm. The Sea-Me quotes a 34 square meter radar cross section equivalent. Another advantage over passive reflectors is that they are more effective under heel angles.

Search and Rescue radar transponders (SARTs)

Under GMDSS the SART units are required on all vessels over 300 GRT. These devices are designed for use in search and rescue, and are different from RTEs. An EPIRB will put potential rescue vessels in the area, but the transponder will accurately localize your position to search radars. Small boat SARTs operate when interrogated by an X-band radar and enhance the return signal. Activation of these units is manual. They typically have a 96-hour operating life in receive mode but only 8 hours when operating.

How does a SART work?

The transponder is not unlike an RTE in operation. Units typically have the following characteristics:

1. **Signal Transmission.** The transponder responds automatically and emits a 9200-9500 GHz high-speed frequency sweeping signal which is synchronous with received scanning radar pulse.

2. **Signal Reception.** On reception of the signal, the position is indicated on radar screens as a line of 12 blips giving range and bearing.

3. **Transponder Receiver.** The transponder gives an audible alarm when the radar emission of a search and rescue vessel is detected.

3. NAVTEX

What is NAVTEX?

NAVTEX is an integral part of GMDSS as well as the Worldwide Navigational Warning Service (WWNWS). It is an automated information system providing meteorological, navigation, and maritime safety information (MSI). Messages are broadcast in English on a pre-tuned and dedicated frequency of 518 kHz with an additional dual mode frequency of 490 kHz now being implemented within UK/Europe and worldwide in local languages. Range is typically around 250 nm, sometimes more or less. INMARSAT enhanced group calling (EGC) provides long-range information. Each of the 16 NAVAREA are divided into four groups, each with up to 6 transmitters with an allocation of 10-minute transmissions each four-hour period. This is time shared to prevent interference on adjacent areas, and they have limited power outputs. Message reception requires a dedicated receiver. Broadcast times are included within frequency listings. There are dedicated units with LCD display units, and also some are printer versions. I use the ICS Nav-6 Plus LCD type on my own boat, and I have to say it is great!

Dual Frequency Navtex with LCD Display
Courtesy of ICS Electronics

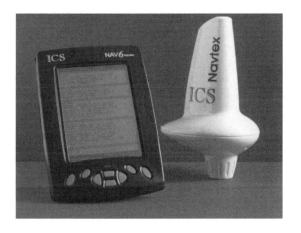

What are message priorities?

Prioritization is used to define message broadcasts. Vital messages are broadcast immediately, usually at the end of any transmission in progress. Those classed as important will be broadcast at the first available period when the frequency is not in use. Routine messages are broadcast at the next scheduled transmission time. The messages classified as vital and important will be repeated if still valid at the following scheduled transmission times. Messages incorporate a subject indicator code (B2 character), which allows acceptance and rejection of specific information. Navigational and meteorological warnings and SAR information are non-selective so that all stations receive important safety information. B2 codes include Nav Warnings (buoy positions altering, wrecks, floating hazards, oil rig moves, naval exercises, meteorological warnings (gales etc), ice reports, SAR and anti-piracy info (cannot be rejected), weather forecasts shipping and synopsis, pilot service messages, Loran messages, OMEGA messages, GPS messages, other NAVAID messages, Nav warnings additional to A (cannot be rejected). (A = Nav warnings, B=Gales, D=Distress information, E=Forecasts.)

What is station identification?

Navigation information is broadcast from a number of stations located within each NAVAREA, and broadcast times as well as transmitter power outputs are carefully designed to avoid interference between stations. Each station is assigned an identification code (B1 character). This is essential so that specific geographical region stations can tune in. Stations are selected by letter designation such as M-Casablanca, I-Las Palmas. The letter Z indicates there are no messages to transit, checks system and is an operational check message. There are two 24 hr daily forecasts for sea areas; the message format contains the following:

> Format nine characters, header code followed by technical code
>
> ZCZC B1 (Transmitter ID) B2 (Subject ID), B3, B4 (Consecutive Number)
>
> Time of origin
>
> Series ID and consecutive number
>
> The message text
>
> NNNN (End of message group)

4. VHF RADIO COMMUNICATIONS

What is VHF radio?

VHF (Very High Frequency) radio is probably the most useful radio communication system available. It allows easy ship-to-ship, or ship-to-shore communications. It permits fast communications with the Coast Guard, harbor masters, other vessels, weather stations, bridge and lock operators and a lot more. The disadvantage is that the communication range is line of sight, typically a maximum of around 35 miles. This means that mountains, headlands, or any other mass can block signal transmission.

About VHF radio theory

The VHF radio spectrum consists of 55 channels in the 156.50 to 163.275 MHz band. As VHF radio is line of sight the higher the two antennas are mounted, the greater the distance. Atmospheric conditions and the installation itself such as aerial type and placement also affect the actual range.

About VHF licensing

All countries have licensing regulations that must be adhered to. Failure to comply may result in prosecution and fines.

1. **Ship Station License.** All VHF installations must possess a station license issued by the appropriate national communications authority, i.e. FCC. On issue of the first license a call sign is issued.

2. **Operator License.** At least one operator, normally the person registering the installation should possess an operator's license or certificate. Under GMDSS and DSC this has changed. In the UK it is the Short Range Certificate (SRC), similarly the requirements have changed in the US, Canada and Australia. It requires a short one-day course.

All about VHF propagation

VHF signals penetrate the ionosphere rather than reflect. In some circumstances VHF signals can reflect back from the ionosphere to give "freak" long distance communications such as during very strong solar cycles. This occurred during cycle 19 in 1957/58, cycle 21 in 1980 and cycle 22 in 1990. During these peaks, the monthly sunspot average rose to extremely high values and the ionosphere reflected higher frequencies than normal. VHF can also be reflected from clouds of increased ionization in the E layer of the ionosphere, and during auroras, which are the light curtains caused by charged particles from the sun.

VHF power consumption

Typical units consume 5-6 amps when transmitting. Reception-only consumption can add up as the set is on for virtually 24 hours. This is in the range of 1 to 7 amps. In a day, that can add up to 12-17 amp-hours depending on the radio set. VHF is however one piece of equipment that should be left on regardless of power consumption. The merchant ship that sights you and tries to communicate will do so well before you may be aware of it.

About VHF talk technique

Hold the microphone approximately 2 inches from the mouth and speak at a volume only slightly louder than normal. Be clear and concise and don't waste words. Many newer sets also incorporate noise-cancelling microphones, which are a good development.

About the VHF radio features

As VHF radio is widely used by official and commercial operators, it is essential to use your set properly for optimum performance.

1. **Power Setting.** Always use the 1-watt low-power setting for local communications, and the 25-watt high power for distance contacts. Some radios automatically power down to 1 watt for Channels 13 and 67.

2. **Squelch Setting.** Squelch reduces the inherent noise in the radio, but do not reduce the squelch too far.

3. **Simplex and Duplex.** Simplex means that talk is carried out on one frequency. Duplex is where transmit and receive are on two separate frequencies.

4. **Dual and Tri-Watch.** This facility enables continuous monitoring on Channel 16 and the selected channel. Tri-watch allows monitoring of a selected channel, Channel 16 and Channel 9 for USCG.

5. **All Channel Scanning.** This makes the radio listen to each channel sequentially for traffic or activity.

6. **Programmable Scanning.** This allows you to store certain channels into memory and you can scan and listen to important channels.

7. **Weather Alert.** The radio will alert with a special tone when any urgent NOAA weather broadcasts are being transmitted. Some radios will automatically tune to the broadcast.

About handheld VHF radios

Many boats, including mine, also carry handheld VHF radios. They provide very easy communications back to the boat when in the dinghy, ashore or working on board. They are also ideal for smaller trailerboats. Most handheld VHF radios are not part of the GMDSS system. Simrad have a handheld radio with built-in DSC distress alert facility. The handheld VHF has a power output limited to 6 watts unlike the larger radios with a 25-watt rating. Typical range for a handheld VHF radio is around 3-8 miles. Battery power is also limited with rechargeable battery packs in the range 500 mAh up to 1400 mAh with power consumption on transmit being around 2 amps. This means that the battery pack soon runs down, so either a spare pack is required or conversation should be limited.

About batteries and waterproofing

Battery lives have been much enhanced with new technologies, such as the introduction of nickel metal hydride (NiMH) and lithium ion (Li-Ion) types. These can give up to 20 hours and, unlike Ni-Cad battery packs, do not suffer from memory effects. Some radios claim to be splashproof and some waterproof, but tests have shown that they do not always perform well in keeping out water. Investing in a waterproof VHF bag is a good idea if your handheld is important in insuring your safety.

About proper VHF radio procedure

After selecting the required channel use the following operating procedures.

1. Wait until any current call in progress is terminated. Even if you do not hear speech, listen for dial tones or other signals. Do not attempt to cut in or talk over conversations. Sometimes traffic may be busy and patience is required.

2. Always identify your vessel and call sign both at the beginning and end of transmission.

3. Keep conversations to a minimum, ideally less than three minutes.

4. After contact with other vessels, allow at least ten minutes before contacting them again.

5. Always observe a three-minute silence period on the hour and half hour. While it is not essential it is good practice.

Making coast station calls

Operate your transmitter for at least 7-8 seconds when calling and use the following format:

1. Call the coast station 3 times.

2. "This is <vessel name and call sign>" and repeat three (3) times.

3. Response will be "Vessel calling <station name> this is <station name> on Channel <No>". This is usually on VHF 16 or the nominated call channel.

4. Response "This is <call sign> my vessel name is <name>."

5. State purpose of business, link call, request for information or advice. "Good evening Sir, I wish to make a transfer charge call". "The number I require is (number)."

6. On completion of business, "Thank you <station> this is <vessel name> over and out, and listening on Channel 16 or <No>."

All about distress, safety and urgency calls

Channel 16 should only be used for the following.

1. **Mayday.** Use of this distress call should only be under the direst of circumstances, "grave and imminent danger." Use of the call imposes a general radio silence on Channel 16 until the emergency is over. Use the following procedure, and allow time before repeating:

   a. "MAYDAY, MAYDAY, MAYDAY."

   b. "This is the vessel <name>."

   c. "MAYDAY, vessel <name>."

   d. "My position is <latitude and longitude, true bearing and distance from known point>."

   e. State <nature of distress> calmly, clearly and concisely.

   f. State type of assistance required.

   g. Additional relevant information including number of people on board.

2. **Pan-Pan.** (Pronounced PAHN-PAHN) Use of this call is to advise of an urgent message regarding the immediate safety of the vessel or crewmember. It takes priority over all traffic except Mayday calls. The call is used primarily in cases of injury or serious illness, or man overboard:

   a. "<All ships>."

   b. "PAN PAN, PAN PAN, PAN PAN."

   c. "This is the vessel <name>."

   d. Await response and transfer to working channel.

3. **Security.** (Pronounced SAY-CURE-E-TAY). For navigational hazards, gale warnings, etc as follows:

   a. "SAY-CURE-E-TAY, SAY-CURE-E-TAY, SAY-CURE-E-TAY."

   b. "This is the vessel/station <name>>."

   c. Pass the safety message.

4. **Medical Services.** Use of this call is to advise of an urgent medical emergency. It takes priority over all traffic except Mayday calls.

   a. "PAN PAN, PAN PAN, PAN PAN."

   b. "RADIOMEDICAL or MEDICO."

   c. "This is the vessel <name, call sign, nationality>."

   d. "My position is <latitude and longitude.> Diverting to <location>."

   e. Give patient details, name, age, sex, and medical history. Give present symptoms, advice required, and medication on board.

The phonetic alphabet

A. ALFA	N. NOVEMBER
B. BRAVO	O. OSCAR
C. CHARLIE	P. PAPA
D. DELTA	Q. QUEBEC
E. ECHO	R. ROMEO
F. FOXTROT	S. SIERRA
G. GOLF	T. TANGO
H. HOTEL	U. UNIFORM
I. INDIA	V. VICTOR
J. JULIETT	W. WHISKEY
K. KILO	X. X-RAY
L. LIMA	Y. YANKEE
M. MIKE	Z. ZULU

Phonetic numbers

1. WUN	6. SIX
2. TOO	7. SEVEN
3. THUH-REE	8. AIT
4. FO-WER	9. NINER
5. FI-YIV	0. ZERO

All about VHF aerial performance

The term gain in decibels (dB) is used when describing the rating of VHF aerials. This also describes the increase or gain in transmission power with respect to concentration or focussing of radio energy. An antenna with a high dB rating will concentrate radio energy perpendicular to the shaft of the antenna in a field with a disk-like shape. This concentrates the radio energy so that the radio signal is not wasted either above or below the antenna, thus increasing the range. A high gain antenna has a greater range, but tends to cause fading when the vessel is rolling and pitching. As the gain of an antenna increases the height of the antenna also increases, but the horizontal angle decreases. The lower gain antenna of 3 dB is more reliable with a broader radiation pattern and has greater range under heavy pitch and roll conditions. To illustrate these changes consider that a 3dB antenna gain at 5 feet has a horizontal angle of 80 degrees. A 6dB antenna gain at 8 feet has a horizontal angle of 35 degrees. A 9dB antenna gain at 20 feet has a horizontal angle of 20 degrees.

VHF antenna radiation and aerials

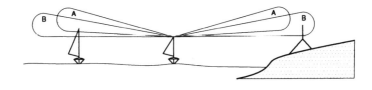

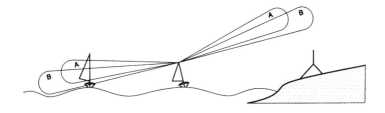

More about VHF aerials

The majority of motorboats use whip aerials and sailing yachts often use low windage stainless steel masthead whips. Trawler yachts and motor sailing vessels can use a masthead-mounted aerial to achieve maximum height. The aerial length is directly related to the aerial gain, and the higher the gain, the narrower the transmission beam. Half-wave whip aerials are typified by the stainless steel rod construction. The radiation pattern has a large vertical component, which suits boats under heel conditions. These antennas can also come in the form of a whip with lengths varying between 1 to 3 meters. The fiberglass whip effectively increases the height, and therefore the range of the radiating element. The gain is typically 3 dB. Increasing height or extending the aerial does not increase gain, but range will improve. Helical aerials have a gain slightly less at 2.5 dB, but do have a characteristically wider signal beamwidth. The higher the gain the more directional the emitted signal becomes. Motorboats usually have 6 dB aerials as they are more directional. The vessels and the aerials are normally vertical as there is roll and pitching but not a permanent heel as in sailing boats.

VHF aerial cables and connections

Cables and bad connections are the principal causes of degraded performance. Avoid using thin RG58U coaxial cable where possible as the attenuation is increased and large signal loss can occur. The amount of signal that is transmitted depends on low losses within the cable and the connections. For cabling aerials always use RG213/U or RG8/U 50 ohms for minimum attenuation. Ensure that the cable has no sharp bends that may affect the attenuation of the cable. The typical cable attenuation of both types for a 100-foot run is:

1. RG58/U. This is a nominal loss of 6.1 dB, which is approximately a percentage signal loss of 75-80%. This means that you will lose around 3dB for around 50 feet of cable.

2. RG8/U and RG8/X (RG213/U). This is a nominal 2.6 dB loss for 100-foot run, which is approximately a percentage signal loss of 45%. The cable adds some weight to the mast, but I personally think that the loss of sailing performance and overall addition to heel is overstated in most sailing vessels.

VHF mast connections

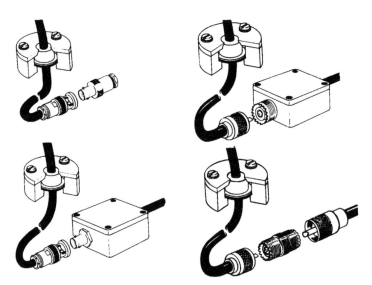

What VHF installation testing is required?

Many vessel VHF installations operate poorly, often with un-diagnosed problems. Many boaters install their own cables, connectors, and aerials, but in the majority of cases the installation is never tested. If the maximum range is to be realized, then the installation requires proper testing. With the increasing reliance on new technology, in particular, with DSC VHF units, reliability and performance are of crucial importance.

Standing Wave Ratio (SWR) checks

When a signal is transmitted via the cable and aerial, a portion of that signal energy will be reflected back to the transmitter. The effect is that coverage is reduced due to the reduced power output. Measure the SWR with a meter. Up until recently you had to hire a technician to bring along an expensive meter (I am fortunate to possess a Bird meter), but you can now use a Shakespeare meter, the ART-3 to check VHF radio performance. This allows easy fault diagnosis and timely repairs, and is highly recommended. A number of problems can reduce the VSWR. Regular testing of reflected power and detection of excessive values will alert you to potential installation problems. It may even save your life.

Common VHF aerial problems

1. **Damaged or Cut Ground Shields.** This is common where the cable has been jointed, or improperly terminated at the connector. Make sure the shield is properly prepared and installed.

2. **Dielectric Faults.** This common problem occurs when cables are run tightly around corners, through bulkheads, and through cable glands. Make sure that cables are bent with a relatively large radius. The tighter the bend, the more dielectric narrowing will occur with increased reflected power.

3. **Pinched Cable.** This common problem occurs where a cable has not been properly passed through a bulkhead with the gland or connector impinging on the cable and reducing its dielectric diameter. Radio waves pass along the outside of the central core and along the inner side of the braiding, so any deformation will alter the inductance and reduce power output.

4. **Connector Faults.** The most common problem is that of connectors not being installed or assembled correctly. Ensure that connectors are properly tightened, that pins are properly inserted, and that the pin-to-cable solder joint is a sound and not a dry joint. Make sure that shield seals are properly made. Many connectors appear good at time of assembly, but deteriorate very quickly when exposed to rain, salt spray, and corrosion. Check the status with a multimeter between the core and screen for short circuits. Consider covering connectors with heat shrink or self-bonding tapes. Coat connections with dielectric silicone compounds to limit corrosion.

5. **Antenna Faults.** If an antenna is out of specification or suffered storm damage, or if a new antenna has been damaged in transit, then functional efficiency will decrease and losses increase. Inspect the antenna and connectors regularly.

United States VHF Channel Designations

Channel	Channel Designation
01	Harbor - Ship-to-Ship
02	Harbor - Ship-to-Ship
03	Harbor - Ship-to-Ship
04	Harbor - Ship-to-Ship
05	Harbor - Ship-to-Ship
06	**SAFETY-SAR Communications - Ship-to-Ship**
07	Commercial Ship-to-Ship
08	Ship-to-Ship Commercial
09	**US Calling Channel (Ship-to-Ship)**
10	Commercial Ship-to-Ship
11	Harbor - Ship-to-Ship
12	Port Operations, traffic advisory, USCG Coast Stations
13	**Bridge and Locks, Ship-to-Ship (1 watt only) Intracostal Waterway (ICW)** Commercial Vessels. Do not use call signs, abbreviated operating procedures only. Maintain dual watch 13 and 16
14	Port Operations - Bridge and Lock Tenders
16	**DISTRESS, SAFETY and CALLING**
18	Commercial Ship-to-Ship and Harbor
19	Commercial Ship-to-Ship and Harbor
20	Port Operations (Duplex)
22	**USCG and Marine Information Broadcasts**
24	Public Telephone - Marine Operator (Duplex)
25	Public Telephone - Marine Operator (Duplex)
26	Public Telephone - Marine Operator (First priority) (Duplex)

27	Public Telephone - Marine Operator (First priority) (Duplex)
28	Public Telephone - Marine Operator (First priority) (Duplex)
60-62	Harbor, Public (Duplex)
63	Harbor, Ship-to-Ship
64	Harbor, Public (Duplex)
65	Port Operations Ship-to-Ship
66	Port Operations Ship-to-Ship
67	(1 watt only)
68	Ship-to-Ship and Harbor
69	Ship-to-Ship and Harbor
70	**DIGITAL SELECTIVE CALLING ONLY (DSC)**
71	Ship-to-Ship and Harbor
72	Ship-to-Ship (Non-commercial)
73	Port Operations
74	Port Operations
77	Ship-to-Ship
78,79,80	Ship-to-Ship and Harbor
81-83	USCG Auxiliary
84	Harbor - Ship-to-Ship - Public telephone (Duplex)
85	Public Telephone (Duplex)
86	Public Telephone (Duplex)
87	Public Telephone (Duplex)
88	Ship-to-Ship
WX1-10	**NOAA Weather broadcasts - Receive Only**

International VHF Channel Designations

Channel	Channel Designation
1, 2, 3, 4, 5, 7	Public Correspondence, Port Operations, Ship-to-Ship, Movement
6	**SAR,** Ship-to-Ship, Movement, Public, Port Operations
8	Public Correspondence, Port Operations, Ship-to-Ship, Movement
9	Public Correspondence, Port Operations, Ship-to-Ship, Movement
10	**SAR,** Ship-to-Ship, Movement, Public, Port Operations
11, 12	Public Correspondence, Port Operations, Ship-to-Ship, Movement
13	**Navigation Safety Communications,** Ship-to-Ship
14	Public Correspondence, Port Operations, Ship Movement
15	Public Correspondence, Port Operations, Ship Movement
16	**DISTRESS, SAFETY and CALLING**
17	Public Correspondence, Port Operations, Ship Movement
18 to 22	Public Correspondence, Port Operations, Ship Movement
23	Public Correspondence, Port Operations, Ship Movement
24	Public Correspondence, Ship Movement
25 to 28	Public Correspondence
60 to 66	Public Correspondence
67	**SAR,** Ship-to-Ship

68, 69	Ship Movement, Ship-to-Ship
70	**DIGITAL SELECTIVE CALLING ONLY (DSC)**
71, 74	Port Operations, Ship Movement
72, 77	Ship-to-Ship
73	**SAR,** Ship-to-Ship, Port Operations
78, 81	Public Correspondence, Port Operations
79, 80	Ship movement, Public Correspondence, Port Operations
82	Public Correspondence
83	Public Correspondence
84	Public Correspondence
85 to 88	Public Correspondence

About VHF networks and frequencies

Rapid growth in cellular phone use has significantly reduced link call activity resulting in the closure of many coast stations. In the US the Maritel Company has bought and opened a private network, and in the future there will be substantial US coastal VHF coverage with automated link call capabilities. This has been a reality for many years in Australia. In Europe Channel 06 is for intership business, Channel 77 for intership chat only. Where the boat is navigating within vessel traffic system (VTS) zones ensure you have the correct frequencies for contacting control stations. The following are public correspondence channels and weather broadcast information. Check regularly for changes and updates as these frequencies may change.

US GREAT LAKES – CANADA ST LAWRENCE.

The Great Lakes and approaches operate on various frequencies. Call on Channel 16. Channels 24, 26, 27, 28 and 85 are the most often used. Traffic, harbor, port and bridge control usually 11, 12, 13 and Channel 14 for locks. Weather is broadcast continuously on WX1, WX2, WX3. **Buffalo** (Weather 22 0255, 1455); **Rochester** 25, 26; **Ripley** 17, 84, 86; **South Amherst/Lorain** (Weather 17 0002, 1102, 1702, 2302); **Erie** 25; **Cleveland** 28, 86, 87; **Toledo** 17, 25, 84, 87; **Detroit** 26, 28 (Weather 22 0135, 1335); **Port Huron** 25; **Grand Haven** (Weather 22 0235, 1435); **Harbor Beach** 17, 86, 87; **Bath City** 28; **Spruce** 17, 84, 87; **Frankfort** 28; **Milwaukee** (Weather 22 0255, 1455); **Michigan City** 25; **Chicago** 26, 27; **Port Washington** 17, 85, 87; **Sturgeon Bay** 28; **Hessel** 17, 84, 86; **Sault Sainte Marie** 26 (Weather 22 0005, 1205); **Grand Marais** 28; **Marquette** 28; **Copper Harbor** 86, 87; **Duluth** 28. **St Lawrence:** most stations 16 and primary channels 24, 26 or 27.

GULF OF MEXICO

WLO on Ch 25, 28, 84, 87 with traffic lists on the hour. **St Petersburg** (Weather 22 1300, 2300); **Marathon** 24; **Naples** 25; **Cape Coral** 26; **Venice** 28; **Palmetto** 25, 27; **Tampa Bay** 86; **Clearwater** 24, 26; **Crystal River** 28; **Cedar Key** 26; **Panama City** 26; **Mobile** (Weather 22 1020, 12220, 1620, 2220); **Pensacola** 26; **Pascagoula** 27; **Gulfport** 28; **New Orleans** 24, 26, 27, 87 (Weather 22 1035, 1235, 1635, 2235); **Venice** 24, 27, 28, 86; **Leeville** 25, 85; **Houma** 28, 86; **Morgan City** 24, 26; **Lake Charles** 28, 84; **Port Arthur** 26, 27; **Galveston** 25, 86, 87 (Weather 22 1050, 1250, 1650, 2250); **Port Lavaca** 26, 85; **Corpus Christi** 26, 28 (Weather 22 1040, 1240, 1640, 2240); **South Padre Island** 26.

ATLANTIC COAST

Southwest Harbor 28 (Weather 22 1135, 2335); **Camden** 26, 27, 84; **Portland** 24, 28 (Nav 22 1105, 2305); **New Hampshire** 28; **Gloucester** 25; **Boston** 26, 27 (Weather 22 1035, 2235); **Hyannis** 28, 84; **Nantucket** 27, 85, 86; **Woods Hole** (Weather 22 1005, 2205); **New Bedford** 24, 26, 87; **Providence** 27, 28; **Bridgeport** 27; **Riverhead** 28; **Long Island Sound** (Weather 22 1120, 2320); **Bay Shore** 85; **Moriches** (Weather 22 0020, 1220); **New York** 25, 26, 84 (Weather 22 1050, 2250); **Sandy Hook** 24 (Weather 22 1020, 2220); **Bayville** 27; **Atlantic City** 26; **Cape May** (Weather 22 1103, 2303); **Philadelphia** 26; **Wilmington** 28; **Dover** 84; **Delaware Bay** 27; **Salisbury** 86; **Cambridge** 28; **Baltimore** 24 (Nav 22 0130, 1205); **Point Lookout** 26; **Norfolk** 25, 26, 27, 84; **Hampton** 25, 26, 27, 84; **Georgetown** 24; **Charleston** 26 (Weather 22 1200, 2200); **Savannah** 27, 28; **Brunswick** 24; **Jacksonville** 26; **Daytona Beach** 28; **Cocoa** 26; **Vero Beach** 27; **West Palm Beach** 28, 85; **Boca Raton** 84; **Fort Lauderdale** 26, 84; **Miami** 24, 25 (Weather 22 1230, 2230); **Miami Beach** 85; **Homestead** 27, 28; **Key West** 24, 84 (Weather 22 1200, 2200).

PACIFIC COAST (CANADA–CG)

Vancouver 26; Van Inlet, Barry Inlet, Rose Inlet, Holberg, Port Hardy, Alert Bay, Eliza Dome, Cape Lazo, Watts Point, Lulu Island, Mount Parke, Port Alberni 26; Dundas, Klemtu, Cumshewa, Naden Harbour, Calvert, Nootka, Mount Helmcken, Mount Newton, Bowen Island, Texada, Discovery Mt 84. Weather continuous broadcast on WX1, WX2, WX3, 21B.

PACIFIC COAST (US)

Bellingham 28, 85; Camano Island 24; Seattle 25, 26 (Weather 22 CG 0630, 1830); Tacoma 28; Cosmopolis 28; Astoria 24, 26 (Nav 22 CG 053, 1733); Portland (Nav 22 CG 1745); Newport 28; Coos Bay 25; Brookings 27; Humboldt Bay (Weather 22 CG 1615, 2315); Casper 28; Point Reyes 25; San Francisco 26, 84, 87; Santa Cruz 27; Monterey 28 (Weather 22 CG 1615, 2345); Long Beach (Weather 22 CG 0203, 1803); Santa Barbara 22, 86; Avalon/San Pedro 24, 26; San Diego 28, 86 (Nav 22 CG 0103, 1703).

CARIBBEAN

1. MEXICO. Chetumal 26, 26; Cozumel 26, 27; Cancun 26, 27; Veracruz 26, 27. Channel 68 is channel for local cruiser nets.

2. BERMUDA. 27, 28. Coastal forecasts on Channels 10, 12, 16, 27, 38 at 1235 and 2035.

3. BAHAMAS. Nassau 16, 27; 8 Mile Rock 27; Exuma 22 CG; Marsh Harbor 16. Forecasts every odd hour Channel 27. Cruisers Net operates on Channel 68 at 0815 with weather forecasts etc.

4. CAYMAN ISLANDS. Radio Cayman 1205, 89.9, 105.3 at 0320, 1130, 1220, 1230, 1330, 1710, 2320.

5. JAMAICA. Kingston 16, 26, 27. Forecasts for SW, NW, and Eastern Caribbean, and Jamaica coastal waters forecast at 0130, 1430, and 1900 on Channel 13.

6. PUERTO RICO (USCG). Santurce 16, 26. NOAA forecasts broadcast continuously on VHF WX2 and VHF 22 at 1210 and 2210.

7. VIRGIN ISLANDS (US). St Thomas 16, 24, 25, 28, 84, 85, 87, 88. Forecast West North Atlantic, Caribbean and Gulf of Mexico on Ch 28 at 0000 and 1200. Channels 16, 24, 25, 28 (Traffic Lists), 84, 85, 87 and 88.

8. VIRGIN ISLANDS (UK). Tortola 16, 27. Weather on ZBVI Radio 780 at 0805, and every H+30 0730-1630, 1830-2130 LT.

9. WINDWARD ISLANDS. Martinique (Fort-de-France), Guadeloupe 16, 11. Warnings on receipt odd H+33 and VHF 26 and 27 every odd H+30. Weather messages VHF 26 and 27 at 0330 and 1430.

10. BARBADOS. 16, 26. Forecasts at 0050, 1250, 1650, 2050. Warnings on receipt and every 4 hrs for Caribbean, Antilles, Atlantic waters on Channel 26.

11. GRENADA. St George's 16, 06, 11, 12, 13, 22A. Forecast on request.

12. TRINIDAD and TOBAGO. 16, 24, 25, 26, 27. Forecast at 1340 and 2040.

EUROPE, UK, MEDITERRANEAN

United Kingdom. MSI forecasts 10, 23, 73, 84 and 86. Local Inshore Forecasts, strong wind warnings every 4 hours start with first time listed, area forecasts every 12 hours start time in brackets. Swansea 0005 (0805) Thames 0010 (0810) Clyde 0020 (0820) Yarmouth 0040 (0840) Solent 0040 (0840) Brixham 0050 (0850) Dover 0105 (0905) Shetland 0105 (0905) Stornaway 0110 (0910) Falmouth 0140 (0940) Forth 0205 (1005)

Liverpool 0210 (1010) **Portland** 0220 (1020) **Holyhead** 0235 (0635) **Belfast** 0305 (0705) **Aberdeen** 0320 (0720) **Milford Haven** 0335 (0735) **Humber** 0340 (0740)

NETHERLANDS

VHF 13 is for intership communications. 24 hour watch on **Schiermonnikoog** 5; **Brandaris** 5; **Den Helder** 12; **IJmuiden** 88; **Scheveningen** 21; **Hoek van Holland** 1, 3; **Ouddorp** 74; **Vlissingen** 14, 64. Weather forecasts for Dutch coastal waters and IJsselmeer at 08.05, 13.05 and 23.05 local time. (Local Time is UTC + 2 hours March to October and UTC +1 October to March). Gale warnings and safety messages are on Channels 23 and 83, after prior announcements on Channel 16. Scheduled broadcast times are at 03.33, 07.33, 11.33, 15.33, 19.33 and 23.33 UTC.

IRELAND

Forecast on **Valentia** at 0103 and every 3 hours to 2203 on Channel 24. **Malin Head** on Channel 23 for **Fastnet, Shannon** and Irish coastal waters. **Valentia** MF on ITU Ch 278 and 280, **Malin Head** on ITU Channel 244 and Channel 255.

CHANNEL ISLANDS

Jersey. Weather at 0645, 0745, 1245, 1845, 2245 on Channels 25, 82.

FRANCE

Weather Bulletins all in French. Storm, gale and nav warnings (times in brackets) in English and French. **Gris-Nez** 79 (H+03); **Jobourg** 80 (H+03); **Corsen** 79 (H+03): **Etel** 80 (H+03); **Soulac** 16, 15, 67, 68, 73; **Agde, LaGarde, Corsica** 16, 11, 67, 68, 73; **Monaco** 16, 20, 22, 23, 86. Weather 0903, 1403, 1915.

SPAIN

Weather channel and time in brackets. **Bilbao** 26; **Santander** 24 (Ch11 @ 0245, 0645, 1045, 1445, 1845, 2245; **Cabo Peñas** 26; **Coruña** 26 (Ch26 @ 0803, 0833, 2003, 2033); **Finisterre** 01, 02 (Ch11 @ 0233 every 4 hrs); **Cádiz** 26; **Tarifa** 82 (Ch10, 74 @ 0900, 2100); **Málaga** 26; **Cabo Gata** 27; **Almeria** (Ch10, 74 @ every H+15); **Cartagena** 04; **Alicante** 0; **Valencia** (Ch10 @ every even Hr +15); **Algeciras** (Ch15, 74 @ 0315, 0515, 0715, 1115, 1515, 1915, 2315; **Ibiza** 03; **Palma** 07; **Menorca** 87; **Las Palmas.** Weather Forecast on 04, 05, 26, 28 at 0903, 1203 and 1803. **Arrecife** 25; **Fuerteventura** 22, 64; **Tenerife** 27; **La Palma** 22.

PORTUGAL

Arga 25, 28, 83; **Arestal** 24, 26, 85; **Monsanto** (Channel 11 @ 0250, 0650, 1050, 1450, 1850, 2250); **Montejunto** 23, 27, 87; **Lisboa** 23, 25, 26, 27, 28; **Atalaia** 24, 26, 85; **Picos** 23, 27, 85; **Estoi** 24, 28, 86; **Sagres** (Channel 11 @ 0835. 2035); **Acores.** Forecast on 16, 23, 26, 27, 28 at 0935 and 2135. **Madeira** 25, 26, 27, 28.

ITALY

(Sardinia) **Monte Serpeddi** 04; **Margine Rosso** 62; **Porto Cervo** 26; (West Coast) **Monte Bignone** 07; **Castellaccio** 25; **Zoagli** 27; **Gorgona** 26; **Monte Argentario** 01; **Monte Cavo** 25; (West Coast/Sicily) **Posillipo** 01; **Capri** 27; **Sera del Tuono** 25; **Forte Spuria** 88; **Cefal** 61; **Ustica** 84; **Erice** 81; **Pantelleria** 88; **Mazarra del Vallo** 25; **Gela** 26; **Siracusa** 85; **Campo Lato Altoi** 86; **Lampedusa Ponente** 25; **Crecale** 87; **Capa Armi** 62; **Ponta Stilo** 84; **Capo Colonna** 88; **Monte Parano** 26; **Abate Argento** 05; **Bari** 27; **Monte Calvario** 01; (Adriatic) **Silivi** 65; **Monte Secco** 87; **Forte Garibaldi** 25; **Ravenna** 27; **Monte Cero** 26; **Piancavallo** 01; **Conconello** 83. (Weather bulletins on Channel 68 @ 0135, 0735, 1335, 1935; nav and gale warnings on receipt and H+03 and H+33, continuous broadcast in Northern Adriatic).

GREECE

Kerkyra (02) 03, 64; Kefallinia 26, (27), 28; Koryfu 87; Petalidi 23, (83), 84; Kythira (85), 86; Poros 27, 28, 88; Gerania 02, 64; Perama (Piraeus) 25, 26, 86, 87; Parnis (25), 61, 62; Lichada 01; Pilio 03, (60) ; Sfendami (23), 24; Tsoukalas 26, 27; Thasos 25, 85; Limnos (82), 83; Mytilini (01), 02; Chios 85; Andros 24; Syros 03, (04) ; Patmos 84; Milos 82; Thira 26, 87; Kythira 85, 86; Astypalea 23; Rodos 01, (63) ; Karpathos 03; Sitia (85), 86; Faistos 26, 27; Moystakos 04; Knossos (83), 84. Hellas Channels in brackets at 0600, 1000, 1600, 2200. Hellas Channel 86 has Wx bulletin from Perama.

TURKEY

Akcakoca 01, 23; Keltepe 02, 24, 82; Sarkoy 05, 27; Camlica 03, 07, 25, 28; Mahyadagi 04, 26; Kayalidag 01, 23; Akdag 02, 24, 28; Izmir 16, 04, 24; Antalya 25, 27; Dilektepe 03, 07, 25; Palamut 04, 05, 26; Yumrutepe 01, 23; Anamur 03, 25; Cobandede 02, 26; Markiz 02, 24.

CYPRUS

Olympos 16, 26, 24, 25, 26; Kionia, Pissouri, Lara 25, 26, 27.

MALTA

01, 02, 03, 04, 16, 28. Traffic Lists on 04.

CROATIA

Senj, Pula, Zadar 10, 16; Rijeka 04, 16, 20, (24) 0535, 1435, 1935; Split 16, 21 (0545, 1245, 1945); Dubrovnik 07, 63, (04) (0625, 1320, 2120); Channels 67, 69, and 73 have continuous weather forecasts for Northern and Central Adriatic Sea updated three times per day in English.

AUSTRALIAN AND NEW ZEALAND VHF CHANNELS

VHF coast stations and supplementary safety channels. All stations monitor 16 and 67.

AUSTRALIA

QUEENSLAND. Seaphone and Weather (0633 1633 EST) Weipa 03 Torres Strait (Mia Is) 26 Thursday Is. 66 Torres Strait (Darnley Is) 60 Lochhart River 28/26 Cooktown 61 Cairns 27/24 Townsville 26/23 Ayr/Home Hill 60 Whitsunday Is. 25/28/83/86 Shute Harbour 66 Mackay 65 (0733 1803 EST) Port Clinton 01/04 Yeppoon 61 Gladstone 27/24 Fraser Island 62 Sunshine Coast 25/28 Brisbane Central 02 Gold Coast 23/26.

NEW SOUTH WALES Seaphone and Weather (0648 1818 EST) Coffs Harbour 27 Camden Haven 62 Port Stephens 25/28 Lake Macquarie 01, Hawkesbury River 02/05 Sydney 23/26/63 Sydney South 84 Nowra 27 Eden 86.

VICTORIA Seaphone and Weather (0803 1733 EST) Lakes Entrance 27 Port Welshpool 60 Melbourne 23/26.

TASMANIA Seaphone and Weather (0803 1733 EST) North Tasmania 28 Hobart 07 Bruny Island 24/27 St Marys 26.

SOUTH AUSTRALIA Seaphone and Weather (0748 1718 CST) Adelaide 23/26 Kangaroo Island 61 Port Lincoln 24/27.

WESTERN AUSTRALIA. Seaphone and Weather (0633 1703 WST) Rottnest Island 60 Jurien Bay 62 Geraldton 28 Carnarvon 24 Dampier 26 Port Headland 27 Broome 28.

NORTHERN TERRITORY Seaphone and Weather (0803 1833 CST) Darwin 23/26 Gove 28.

NEW ZEALAND

Channel 16 and Working Channels 25, 67, 68, 69 or 71 Safety Information and Weather at 0533, 0733, 1033, 13333, 1733, 2133. Cape Reinga 68, Kaitaia 71 Whangarei 67 Great Barrier Is 25, 67, 68, 71 Plenty 68 Runaway 71 Tolaga 67 Napier 68 Wairarapa 67 Wellington 71 Picton 68 Kaikoura 67 Akaroa 68 Waitaki 67

Chalmers 71 Bluff 68 Stewart Island 71 Puysegur 67 Fiordland 71 Fox 67 Greymouth 68 Westport 71 Farewell 68 D'Urville 67 Wanganui 69 Cape Egmont 71 Taranaki 71 Auckland 71.

About cell phone communications

The rapid development of the mobile cellular telephone has made personal communications in coastal waters much easier. Many of us are fortunate to be able to utilize GSM technology. I can use my phone offshore and in more than 30 countries. This technology has not been without a price. The rapid drop in placement of link calls has meant the closure of many coast stations and repeaters. I have had some interesting calls: on one occasion I had a call from a motorboat regarding a charging problem. After I enquired about his location, I found out that he was some 2000 miles away, and 10 nm offshore. On another occasion a skipper in a major offshore yacht race called me during the race and described his problem; I was able to nurse him to the finish line. More recently there have been text message SOS calls from Indonesia to the UK setting off SAR via Australia, and people in life rafts calling for help. The cell phone must be put into perspective and it must be emphasized that it is not a substitute for VHF or HF marine communications systems.

Distress calls and cellular phones

There are many reasons why cell phones are not good for distress unless as a last resort. A vessel in distress cannot communicate with other potential rescue vessels in the area. This has the effect of delaying rescues considerably, using greater resources and increasing the risks to all involved. If you are in distress, you simply may put on hold and not get through to an appropriate authority, or you may be at the outside of the cell range and drop out repeatedly. Vessels in distress who cannot provide exact position information cannot be located using VHF direction-finding (DF) equipment. Vessels in distress cannot activate priority distress alerting using cell phones. Rescue scene communications can be severely disrupted because normal cell phone communications can only occur between two parties. Most rescue vessels and SAR aircraft do not have cellular phones. These communications problems and resulting message exchanges have the potential to cause disruptions or delays to the extent that a safe rescue opportunity may be lost with catastrophic results. It is worth noting that a number of system operators have introduced services for vessel cellular phone users. Bell Atlantic in the US offers direct contact with the Coast Guard (press star key followed by "C" and "G).

How good is cell phone coverage?

If you use your cellular phone on coastal trips, you will have a few problems with dropouts. The problems will occur at the outer range of the transmission cell, and this is more pronounced at sea. Range is typically 8 miles, which is a lot less than VHF. If you really need to use a cellular phone regularly, install an external aerial for maximum range. Many lament the passing of analog to digital, as range was better. If you wish to install a set on board it is worth considering buying a higher 8-watt set such as from Motorola. Install it to the vessel power supply and add an external whip aerial to maximize range and power. It is also worth considering a dual or tri-band phone to maximize operational range.

Cell phone SIM card change outs

Global roaming rates are not cheap. If you are in one place for any length of time, and make lots of local calls consider buying a pay-as-you go SIM card and number. This offers real long-term savings in most countries. I have found it the ideal solution in Europe.

About text messaging

For locations in reach of GSM, short text messaging is a useful low-cost way of staying in touch. Make sure you can both send and receive text messages if you intend using this method. I have had problems in some European areas.

5. SSB/HF RADIO COMMUNICATIONS

While SSB (Single Side Band) HF (High Frequency) is part of GMDSS many boaters opt to keep using sets without the capability, in particular for e-mail and weatherfax. Long-range radio communications depend on radio frequencies in the HF spectrum of 2 to 24 MHz. Radio signals consist of a carrier signal and audio sidebands, both upper and lower. To make transmission more effective one side band is used to transmit all of the power giving good communications ranges.

About SSB signal propagation

Skywaves travel up until they reach the ionosphere and are bent and reflected back over a wide area. The ionosphere exists at a height of 30 to 300 miles (50 to 500 kms) above earth. It is formed by the ionization of air atoms by incoming UV ionizing radiation from the sun. The ionosphere is a weakly ionized plasma that is constantly changing and these changes alter the propagation characteristics of the radio waves. This is typified by the differences in night and day time transmission characteristics. Good HF communications depend on utilization of the changing conditions with use of optimum frequencies. The ionosphere structure is divided into layers D, E, F1 and F2 in order of increasing height. A software package, HFProp, is useful at predicting optimum transmission times worldwide.

> **F Layer.** The main reflecting layer is called the F layer and is approximately 200 mile (320 km) high. The layer is permanently ionized but during daylight hours energy from the sun causes the intervening layers E and D to form, and at night these reflect the highest radio frequencies in HF bands.

E and D Layers. The signals reflected from these layers have lower ranges. Frequencies of 3 MHz or less are absorbed by the D layer, and eliminate skywave propagation. For this reason 2 MHz is not favored.

Ground Wave. Ground wave signals travel along the earth's surface but are absorbed or masked by other radio emissions.

Skip Zone. The skip zone is the area between the transmission zone and the zone where the signal returns to earth, with generally negligible signal.

HF Radio Wave Behavior

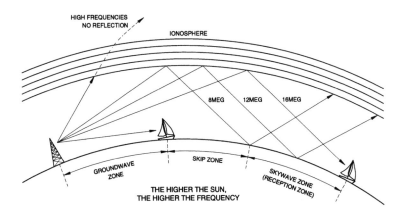

About radio propagation changes

The ionosphere will affect each frequency differently. Extreme Ultra-Violet (EUV) radiation is responsible for forming and maintaining the ionosphere and depends on solar sunspot activity. When sunspot activity is lowest, the solar cycle EUV radiation is also weak and the density of charged particles in the F region is lowest. In this state only lower frequency HF signals can be reflected. At sunspot cycle peaks the EUV and ionosphere density are high and therefore higher frequencies can be reflected. The season, time of day, and the latitude also affect HF radio communications. Solar flares produce high levels of electromagnetic radiation, and the x-ray component increases the D layer ionization. As HF communications use the F layer above they must transit the D layer twice during any signal skips. During a major solar flare the increased ionization results in a higher density of neutral particles and absorption of signal in the D layer. This is called sudden ionosphere disturbance (SID). It is characterized by increased attenuation of HF signals at lower frequencies, and is also called short wave fadeout, SSWF for a sudden fadeout and GSWF for a gradual one. These events are synchronized with solar flare patterns. They are characterized by rapid onsets of just several minutes and declines of up to an hour or greater.

Space weather effects and HF radio

Space weather has become very important in the satellite communications age. The underlying factor ruling space weather, at least in this end of the galaxy, is our sun. The sun is by nature prone to dramatic and violent changes, with events such as solar flares, and the resulting blast streams of radiation and energized particles that stream towards earth. Space weather is caused by changes in the speed or density of the solar wind, and this is the continuous flow of charged particles that flow from the sun past earth. The flow tends to distort the earth's magnetic field, compressing it in the direction of the sun and stretching it out in the opposite direction. The solar wind fluctuations cause variations in the strength and direction of the magnetic field near the earth's surface, and sudden variations are called geomagnetic disturbances. The electrical layers of the ionosphere are disrupted.

How solar cycles can cause problems

Space weather depends on an 11-year solar cycle. Cycles vary in both intensity and length, and the solar activity is characterized by the appearance of sunspots on the sun. Sunspots are regions of stronger magnetic field, and the solar maximum is the time when maximum spot numbers are visible. Sunspot numbers are quoted for average numbers over a 12-month period, and are the traditional measure of solar cycle status. Peak sunspot activity is calculated a little like the highest rainfall and is for recorded worst cases. Five of the last six solar cycles have been of high magnitudes. Cycle 19 in 1957 peak had a sunspot number of 201, the largest on record. Cycle 21 in 1979 had a peak sunspot number of 165 and was the second largest. Cycle 22 in 1989 was the third largest. The present Cycle 23 is due to end in early 2007 after it peaked in 2000 and any major solar disruption should gradually diminish.

Selecting the best frequency preferences

The best ocean frequencies are on 4 MHz with ranges of up to 300 miles in day conditions, and thousands of miles at night without static at 2 Mhz. The principal characteristics are:

1. **Sunset.** At sunset, the lower layer ionization decreases, and the D layer will disappear.

2. **Dusk.** At dusk, range increases on 2 MHz over thousands of miles, almost instantaneously. The interference levels are dramatically reduced.

3. **Night.** The reflecting layer of the ionosphere rises at night, increasing the ranges for 4-6 MHz so lower frequencies are best at night. High frequencies are not good at night.

4. **Day.** Low frequencies are weak during daytime. High frequencies are used in the daytime.

SSB optimum transmission times sunrise to noon

Frequency (MHz)	Sunrise 0600	Noon 1200
22000	Average 100-2000 nm	Good 2000 nm plus
12000	Good 2000 nm plus	Good 2000 nm plus
8000	Good 2000 nm plus	Average 100-2000 nm
6000	Good 2000 nm plus	Average 100-2000 nm
4000	Average 100-2000 nm	Bad 50 nm
2000	Good 2000 nm plus	Bad 50 nm

SSB optimum transmission times sunset to midnight

Frequency (MHz)	Sunset 1800	Midnight 2400
22000	Good 2000 nm plus	Average 100-2000 nm
12000	Good 2000 nm plus	Good 2000 nm plus
8000	Average 100-2000 nm	Good 2000 nm plus
6000	Average 100-2000 nm	Good 2000 nm plus
4000	Bad 50 nm	Good 2000 nm plus
2000	Bad 50 nm	Good 2000 nm plus

SSB/HF radio operation requirements

There are certain legal requirements and operational procedures to observe.

1. **Ship Station Licensing.** Every vessel must have a license issued by the relevant communications authority. Transmitters must also be of a type approved by the appropriate authority. The issued call sign and vessel name must be used with all transmissions.

2. **Operator Licensing.** An operator's certificate is required and a test is given that covers knowledge of distress and safety procedures, and related marine communications matters.

HF radio frequencies and bands

Always consult a current list of radio signals. The UK Admiralty List of Radio Signals (ALRS) is by far the most accurate, so invest in the relevant volume for any world location.

1. **Listen to Station.** If you can hear traffic clearly on the band you will probably have relatively good communications on that band.

2. **Monitor Bands.** Monitor the various bands and channels and determine the best peak period for communications. If the signal strength is good but the channel is busy, use a second channel if available, or wait. Do not tune equipment while a call is in progress.

3. **Station Identification.** Have name, call sign, position, and accounting code ready for the operator if required.

United States SSB weather frequencies

USCG CAMSPAC (Master Station Pacific) Honolulu (NMO) and Point Reyes (NMC) have very good weather transmissions.

US Coast Guard Channels

ITU Channel Number	Receive Frequency	Transmit Frequency
	2182.0	
424	4426.0	4134.0
601	6501.0	6200.0
816	8764.0	8240.0
1205	13089.0	12242.0
1625	17314.0	16432.0

United States, Canada, Caribbean

The U.S. Coast Guard broadcasts National Weather Service off-shore forecasts and storm warnings on 2670 kHz after an initial announcement on 2182 kHz. Visit http://www.navcen.uscg.gov/marcomms

1. Mobile (WLO) www.wloradio.com

   Tx/Rx (kHz)   4077/4369, 4104/4396, 6218/6519, 8264/8788, 8280/8806, 12263/13110, 12305/13152, 16378/17260, 16480/17362, 22108/22804

   ITU Channels   405, 414, 607, 824, 830. 1212, 1226, 1607, 1641, 2237

   Weather and Traffic List Broadcast Times UTC Gulf of Mexico 0400, 1300, 1600, 2200. SW N Atlantic – 0500, 1300, 1700, 2300. Caribbean Sea – 0600, 1300, 1800, 0000. East Pacific – 0300, 1400, 2000, Alaska Offshore – 0800, 1500

2. Seattle (KLB) (Pacific and Alaska) www.wloradio.com

Tx/Rx	4113/4405, 8207/8731/, 12254/13101, 16429/17311
ITU Channels	417, 805, 1209, 1624

3. Chesapeake (NMN)

Frequencies	4125, 6215, 8291,12290
Weather	Forecasts at 0330, 0515, 0930, 1115, 1530, 1715, 2130, 2315

4. New Orleans (NMG)

Frequencies	4125, 6215, 8291, 12290 kHz.
Weather	Forecasts at 0330, 0515, 0930, 1115, 1530, 1715, 2130, 2315

5. Point Reyes (NMC) (MMSI 003669905)

Frequencies	4426, 8764, 13089, 17314, 4125, 6215, 8291, 12290
Weather	0430, 1030, 1630, 2230

6. Kodiak (NOJ)

Frequencies	6501 kHz
Weather	Forecasts at 0203, 1645

7. Honolulu (NMO) (MMSI 003669990)

Frequencies	6501, 8764, 13089 4125, 6215, 8291, and 12290 kHz.
Weather	0005, 0600, 1200, 1800

8. Guam (NRV)

Frequencies	6501, 13089.
Weather	Forecasts at 0330, 0930, 1530, 2130

9. Miami (NMA)

Frequencies	4125, 6215, 8291 and 12290 kHz.
Weather	Forecasts at 0350, 1550

10. Boston (NMF)

Frequencies	4125, 6215, 8291 and 12290 kHz.
Weather	Forecasts at 1035, 2235

CANADA

1. **Prince Rupert (VAJ)**

Frequencies	2054.
Weather	0105, 0705, 1305, 1905.
Navtex (D)	0030, 0430, 08030, 1630, 2030.

2. **Tofino (VAE)**

Frequencies	2054, 4125.
Weather	0050, 0500, 0650, 1250, 1730, 1850, 2330.
Navtex (H)	0110, 0910, 1310, 1710, 2110.

3. **St Lawrence** (St Johns, Halifax, Sydney Placentia, Port Aix Basques, Rivière-au-Renard, S.Anthony)

Frequencies	1514, 2538, 2582. Traffic lists on 2749, 2582.

4. **Fundy (VAR)** (MMSI 003160015)

Frequencies	2182, 2749.
Weather/Nav	0140, 1040, 1248, 1625, 1730, 1948, 2020.
Navtex (U)	0320, 0720, 1120, 1520, 1920, 2320.

5. **Sydney (VCO)**

Frequencies	2182, 2749.
Weather/Nav	0033, 0733, 1433, 1503, 2133.
Navtex (Q)	0255, 0655, 1055, 1455, 1855, 2255.

6. **Rivière-au-Renard (VCG)**

Frequencies	2182, 2598, 2749.
Weather/Nav	0437, 0847, 0937, 1407, 1737.
Navtex (C)	0020, 0420, 0820, 1220, 1620, 2020.

CARIBBEAN

1. **Bermuda (Bermuda Harbor) (MMSI 003100001)**

   Frequencies   2182, 2582 (ITU 410, 603, 817, 1220, 1618).

   Weather   Coastal forecast 1235, 2035 on www.weather.bm.

   Navtex   0010, 0410, 0810, 1210, 1610, 2010.

2. **Bahamas (Nassau)**

   Frequencies   2182, 2522, 2588, 2522/2126.

   Weather   Forecasts every odd hour on 2522, storm and hurricane warnings are issued on receipt. Radio Bahamas 1540/1240/810, and 107.9-MHz broadcast detailed shipping weather reports M-F at 1205 hrs. Daily weather messages/synopsis 0815, 1315 and 1845 hrs.

3. **Jamaica (Kingston)**

   Frequencies   2182, 2587, 2590, 3535 ITU 405, 416, 605, 812, 1224.

   Weather   Coast Guard on 2738 kHz at 1330, 1830 hrs for SW, NW, and Eastern Caribbean, and Jamaica coastal waters forecast. Radio Jamaica on Montego Bay 550/104.5. Weather messages M-F 0015, 0340, 1104, 1235, 1709, and 2004 hrs. Jamaica B.C. on 560/620.700/93.3 MHz, fishing and weather forecast M-F 2248 hrs.

4. Puerto Rico (USCG) San Juan

Frequencies	2182, 2670 (Santurce) 2182, 2530.
Weather	Forecast at 0030, and 1430.
Navtex (R)	0200, 0600, 1000, 1400, 1800, 2200.

5. US Virgin Islands (St Thomas)

Frequencies	2182, 2506/2009 ITU 401, 604, 605, 804, 809, 1201, 1202, 1602, 1603, 2223.
Weather	Forecast West North Atlantic, Caribbean and Gulf of Mexico on 2506 at 0000 and 1200. Also at 1400, 1600, 1800 and 2000 forecasts for Virgin Islands Eastern Caribbean. Virgin Islands radio on VHF 28, 85 at 0600, 1400, 2200. Detailed Caribbean weather reports. WIVI FM 99.5 MHz 0730, 0830, 1530 and 1630.

6. Curaçao (Netherlands Antilles)

Frequencies	2182, 8725.1.
Weather	Forecast at 1305.
Navtex (H)	0110, 0510, 0910, 1310, 1710, 2110.

7. Barbados

Frequencies	2182, 2582,2723, 2805. ITU 407, 816 (Traffic Lists), 825, 1213, and 1640.
Weather	Forecasts at 0050, 1250, 1650, 2050. Warnings on receipt and every 4 hrs for Caribbean, Antilles, and adjacent Atlantic waters. Caribbean Ham Weather Net (8P60M). Broadcasts out of Barbados on 21.400 MHz daily at 1300 hrs. Receives positions 1300-1330. Translates RFI WFs 1330-1400.

8. Martinique (Windward Islands)

Frequencies	2182, 2545.
Weather	Warnings odd H+33. Weather messages 2545 at 1333.

9. Grenada (Windward Islands)

Frequencies	2182, 1040, 3365, 5010, 1508, 7850.
Weather	Forecast at 2100-0215 (1040), 2230-0215 (3365), 2100-2230 (5010), 2100-0215 (15085). GBC Radio on 535 and 15105. Hurricane warnings on receipt and every H+30 after news 0200, 1030, 1130, 1630, 2030, 2230.

10. Trinidad and Tobago

Frequencies	2182, 2735, 2049, 3165.
Weather	1250 and 1850.

11. Caribbean SSB Weather Nets.

Synoptic forecasts and analysis including hurricane information and tracks for all of Caribbean. Times are all UTC. Frequencies 4003 kHz at 1215 to 1230, 8104 kHz at 1230 to 1300, in the hurricane season also 8107 kHz at 2215 to 2245.

English Channel and Atlantic

The following are selected frequencies for navigational warnings, weather forecast and working frequencies.

1. **NETHERLANDS** (CG Radio) (MMSI 002442000)

Frequencies	2182, MF DSC 2187.5, 3673.
Weather	North Sea forecasts at 09.40 and 21.40 hrs UTC on 3673 kHz. Gale warnings are made on receipt. Scheduled broadcast times at 03.33, 07.33, 11.33, 15.33, 19.33 and 23.33 UTC.
Navtex (P)	0230, 0630, 1030, 1430, 1830, 2230.

2. **BELGIUM** (Oostende Radio) (MMSI 0020050480)

Frequencies	2182, 2761.
Weather	Forecast at 0820 and 1720.
Navtex (M)	0200, 0600, 1000, 1400, 1800, 2200 (for Dover Straits).

3. **UNITED KINGDOM**

MSI forecasts made on MF frequencies with announcement 2182. Gale and storm warnings made every 4 hours starting with first time listed, and area forecasts every 12 hours starting time in brackets. **Clyde** 1883 kHz 0020 (0820) **Yarmouth** 1869 kHz 0040 (0840) **Solent** 1641 kHz 0040 (0840) **Shetland** 1770 kHz 0105 (0905) **Stornaway** 1743 kHz 0110 (0910) **Falmouth** 2226 kHz 0140 (0940) **Holyhead** 1880 kHz 0235 (0635) **Aberdeen** 2226 kHz 0320 (0720) **Milford Haven** 1767 kHz 0335 (0735) **Humber** 2226 kHz 0340 (0740)

Navtex

Cullercoats (G) Gale Warnings 0100, 0500 0900, 1300, 1700, 2100 24 Hr Synopsis 0900, 2100

Niton (E) Gale Warnings 0040, 0440, 0840, 1240, 1640, 2040 24 Hr Synopsis 0840, 2040

Portpatrick (O) Gale Warnings 0220, 0620, 1020, 1420, 1820, 2220 24 Hr Synopsis 0620, 1820

4. CHANNEL ISLANDS

Jersey

Frequencies	2182, 1726.
Weather	0645, 0745, 1245, 1845, 2245.

5. IRELAND

Valentia (MMSI 002500200)

Frequencies	2182, 1752.
Weather	0233, 0303, 0633, 0903, 1033, 1433, 1503, 1833, 2103.
Navtex (W)	0340, 0740, 1140, 1540, 1940, 23403.

6. FRANCE

Boulogne-sur-Mer

Frequencies	2182, 1770, 1692, 1694.
Weather	1770 - H + 03 and H + 33 and 1692 @ 0703, 1833.

Gris-Nez (Cross) (MMSI 002275100); Jobourg (Cross) (MMSI 002275200); Corsen (Cross) (MMSI 002275300)

Frequencies	1650, 2182, 2677.

Etel (Cross) (MMSI 002275000); Soulac (Cross) (MMSI 002275010)

Frequencies	2182, 2677.

Brest

Frequencies	1635, 1671, 1876, 2691, 1862.
Weather	1635 - H + 03 and H + 33 and 1671, 1876, 2691, 1862 @ 0733, 1803 and 1671, 1876 @ 0600.

St Nazaire

Frequencies	2182, 1671, 1876, 1722, 2691, 2740.
Weather	0333, 0733, 0803, 1133, 1533, 1833, 1933, 2133.

7. **SPAIN** (Port Operations on Ch 18, 19, 20, 21, 22, 79, 80).

Coruna (MMSI 002241022)

Frequencies	2182, 1698.
Weather	0803, 0833, 1233, 1733 (Bay of Biscay), 2003.
Navtex (D)	0030, 0430, 0830, 1230, 1630, 2030.

Finisterre

Frequencies	2182, 1764.
Weather	0803, 0833, 1203, 1703 (Bay of Biscay), 2033.

Tarifa

Frequencies	2182, 1704.
Weather	0803, 0833, 1233, 1733, 2003.
Navtex (G)	0100, 0500, 0900, 1300, 1700, 2100.
Navtex (X)	0350, 0750, 1150, 1550, 1950, 2350 (Valencia).

Islas Canarias (Las Palmas Radio) (MMSI 002240995)

Frequencies	2182, 1689, 2045, 2048, 2114, 2191. ITU 406, 604.
Weather	Gale warnings and forecast on 1689, 2820, 4372, 6510 at 0903, 1203 and 1803.
Navtex (I)	0120, 0520, 0920, 1320, 1720, 2120.

8. PORTUGAL

Lisboa (MMSI 002630100)

Frequencies	2182, ITU 802, 813, 1203, 1207, 1615, 1632, Traffic lists on 13083 at even H +05.

Apulia

Frequencies	2182, 2657.
Weather	0735, 1535, 2335.

Sagres

Frequencies	2182, 2657.
Weather	0835, 2035.
Navtex (R)	0250, 0650, 1050, 1450, 1850, 2250.

Acores (Faial) Horta

Frequencies	2182, 1663.5, 2657, 2742, 2748, 4434.9/4140.5.
Weather	Warnings and forecast 0935 and 2135.
Navtex (F)	0050, 0450, 0850, 1250, 1650, 2050.

Madeira

Frequencies	2182, 2843, 2657.
Weather	0905, 2105.

Mediterranean radio frequencies and weather forecasts

1. SPAIN

Tarifa	2182, 1704/2129.
Malaga	2182, 1656/2081.
Cabo de Gata	
Frequencies	2182, 1767.
Weather	0803, 0833, 1233, 1733, 2033.

Palma Majorca (Palma Radio) (MMSI 002241005)

Frequencies	2182, 1755.
Weather	0803, 0833, 1203, 1703, 2033.

2. FRANCE

Nice

Frequencies	1350.
Weather	0725, 1850.

Marseille

Frequencies	675.
Weather	0725, 1850.

Monaco (UTC + 1)

Frequencies	2182, 4363, 8728, 13146 (ITU Med - Ch 403, 804, 1224, 1607, 2225; Atlantic 403, 830, 1226, 1628, 2225).
Weather	On receipt and at H+03. Forecast at 0903, 1403, 1915. On 8728 @ 0715 and 1830, 13146 on request, Atlantic bulletin on 8806, 13152, 17232 and 22846 @ 0930. Coastal continuous on 161.750M. Nav info at 0803 and 2103.

3. ITALY

Cagliari

Frequencies	2182, 1722 (traffic list on 2680, 2683).
Weather	0125, 0725, 1325, 1925.
Navtex (T)	0310, 0710, 1110, 1510, 1910, 2310.
Navtex (V)	0330, 0730, 1130, 1530, 1930, 2330 (Augusta) Porto Torres (2719) Genova (1667, 2642, 2722) Livorno (1925, 2591) Civitavecchia (1888, 2710, 3747) Napoli (1675, 2632, 3735) Palermo (1852) Mazara (1883, 2211, 2600) Lampedusa (1876) Augusta (1643, 2628) Crotone (1715, 2663) S.Benedetto (1855).

Messina

Frequencies	2182, 2789.
Weather	0135, 0233, 0633, 0735, 1133, 1335, 1533, 1933, 1935.

Roma

Frequencies	4292, 8520, 13011, 17160.8.
Weather	0348, 0948, 1518, 2118; Fleet Wx 0830, 2030.
Navtex (R)	0250, 0650, 1050, 1450, 1850, 2250.

Bari (Adriatic)

Frequencies	2182, 2579.
Weather	0125, 0725, 1325, 1925.

Ancona (Adriatic)

Frequencies	2182, 2656.
Weather	0148, 0748, 1348, 1948.

Trieste

Frequencies	2182, 2624.
Weather	0848, 1218, 1648, 2048.
Navtex (U)	0320, 0720, 1120, 1520, 1920, 2320.

Venezia

Frequencies	2182, 2698.
Weather	0135, 0403, 0735, 0903, 1303, 1335, 1935, 2103.

4. CROATIA (Adriatic)

Rijeka (MMSI 002387010)

Frequencies	2182, 1641, 1656.

Dubrovnik

Frequencies	160.95M.
Weather	0625, 1320, 2120.

Split (MMSI 002380100)

Frequencies	160.95M.
Weather	0545, 1245, 1945.
Navtex	(O) 0240, 0640, 1040, 1440, 2240.

5. GREECE (Hellas Radio MMSI 002371000)

Kerkyra (MMSI 237673190)

Frequencies	2182, 2830 (Traffic 2607, 2792, 3613).
Weather	Forecasts at 0703, 0903, 1533, 2133.
Navtex (K)	0140, 0540, 0940, 1340, 1740, 2140.

Limnos

Frequencies	2182, 2730.
Weather	Forecasts at 0033, 0633, 1033, 1633.
Navtex (L)	0150, 0550, 0950, 1350, 1750, 2150.

Rodos (MMSI 237673150)

Frequencies	2182, 2624.
Weather	Forecasts at 0703, 0903, 1533, 2133.

Iraklion Kritis (MMSI 237673180)

Frequencies	2182, 2799, 1742, 3640.
Weather	Forecasts at 0703, 0903, 1533, 2133.
Navtex (H)	0110, 0510, 0910, 1310, 1710, 2110.

Athinai

Frequencies	2182, 1695, 1767, 2590, (8743).
Weather	Forecasts at 0703, 0903, (1215), 1533, (2015), 2133.

6. TURKEY

Izmir (MMSI 002716000)

Frequencies	1850, 2182, 2760.
Weather	0333, 0733, 1133, 1533, 1933, 2333.
Navtex (I)	0120, 0520, 0920, 1320, 1720, 2120.

Antalya (MMSI 002713000)

Frequencies	2182, 2187.5, 2670.
Navtex (F)	0050, 0450, 0850, 1250, 1650, 2050.

7. CYPRUS (MMSI 002091000)

Frequencies	2182, 2187.5, 2670, 2700, 3690. ITU 406, 414, 426, 603, 807, 818, 820, 829, 1201, 1208, 1230, 1603.
Weather	2700 kHz @ 0733, 1533.
Navtex (M)	0200, 0600, 1000, 1400, 1800, 2200.

8. MALTA

Frequencies	2182, 2625. ITU 410, 603, 832, 1216, 1233.
Weather	0103, 0603, 1003, 1603, 2103.
Navtex	0220, 0620, 1020, 1420, 1820, 2220.

SOUTH AFRICA (Capetown) DSC: MMSI 006010001

Frequencies	2182, 1764, 4435, 2191, 17338, 22711. ITU 405, 421, 427, 801, 805, 821, 1209, 1221, 1608, 1621, 1633, 2204, 2206, 2221.
Weather	at 1333, 0948, 1748.
Navtex (C)	0020, 0420, 0820, 1220, 1620 and 2020 UTC.

NAMIBIA (Walvis Bay) DSC: MMSI 006010001

Frequencies	2182, 1764, 2191, 2783, 4125, 4357, ITU 401, 602, 801.

New Zealand (Taupo Maritime) MMSI 005120010

Frequencies	2182, 2207, 4125, 6215, 8291, 12290, 16420.
Weather	0133, 0533, 1333, 1733 on 2207, 4146, 6224. Ocean and islands at 0303, 0903, 1503, 2103 on 6224, 12356.

Australian Radio Frequencies and Weather Forecasts

24 hour distress, safety, urgency watches on 4126, 6215, 8291 kHz and navigation warnings on 8176 kHz twice daily. Special announcements on VMC/VMW 5 minutes to every hour (25 minutes after the hour CST)

Charleville VMC (Australian Weather East)

(0700-1800) 4426, 8176, 12365, 16546 (1800-0700) 2201, 6507, 8176, 12365. Coastal waters forecasts and warnings for QLD, NSW, VIC, TAS and SA. Every hour commencing 0000 EST (0030 CST). High seas forecasts and warnings for Northern, NE and SE areas. Every hour commencing 0000 EST (0030 CST).

Wiluna VMW (Australian Weather West)

(0700-1800) 4149, 8113, 12362, 16528 (1800-0700) 2056, 6230, 8113, 12362. Coastal waters forecasts and warnings for QLD Gulf, NT, WA and SA. Every hour commencing 0000 WST (0030 CST). High seas forecasts and warnings for Northern, Western and SE Areas. Every hour commencing 0000 WST (0030 CST)

Penta Comstat (VZX) (http://www.pentacomstat.com.au/)

Frequencies	ITU 429, 608, 802, 1203, 1602, 2243. 4483 (working) and SelCall available on all channels
Weather	NSW coastal waters forecast (Qld border to Gabo Is) 0725. Queensland coastal waters forecast (Cardwell to Coolangatta) and coastal position report sked 0735. Long range position report sked 0800. NSW coastal waters forecast 1625. Queensland coastal waters forecast and Coastal position report sked 1635. Long range position report sked 1700.
E-mail	Station for SailMail Association. www.pca.cc

NEW ZEALAND (Taupo Maritime Radio) MMSI
005120010

Distress and Safety	4207.5, 6312.0, 8414.5, 12557.0, 16804.5
Frequencies	2182, 2207, 4125, 4146, 6215, 6224, 8297, 12290, 12356, 16420, 16531
Weather (NZST)	Coastal warnings and bulletins 0133, 0533, 1333, 1733 on 2182, 4125, 6215. Coastal reports 0803, 1203 2003 on 2182, 4125, 6215. Oceanic warnings at 0303 & 1503 on 6215, 12290, 6224, 12356. At 0333, 133 on 8291, 16420, 8297, 16531. At 1503, 1533, 2103, 2133, 2103, 2133 on 6215, 12290, 6224, 12356. At 0903, 2103 on 6215, 12290, 6224, 12356. At 0933, 2133 on 8291, 16420, 8297, 16531

About HF radio tuner units

The tuner unit function is to match the antenna length to the frequency being used:

1. **Manual.** There are still manual tuner units around although they have been largely phased out by fully synthesized systems with automatic tuner units. These required the matching of the antennas by adjusting tune and load controls using an inbuilt tune meter.

2. **Fully Synthesized Units.** The new synthesized radio sets with automatic tuner units enable non-technically oriented people to communicate easily. Units consist of a full range of ITU EPROM controlled frequencies. The tuner unit essentially consists of inductors and capacitors that are automatically switched in series or parallel with the antenna to achieve the correct tuned length.

All about HF radio aerials

The aerials are crucial to proper performance of the HF radio. On motorboats, the whip is the most practical. The whip generally operates over a wider frequency range than the wire line aerials seen on sailing boats. Manufacturers are Shakespeare in the US and V-Tronics in the UK.

1. **Loaded Whip.** These aerials have loading coils, and are generally very long.

2. **Unloaded Whip.** These whips have a similar performance to long wire backstay aerials. The ATU provides the required aerial length. As the voltage and currents can be significant at the base, it is essential to use high quality insulators and well insulated feed line cables to minimize losses. A very low resistance ground system is required.

Tuner Unit and Aerial Connections

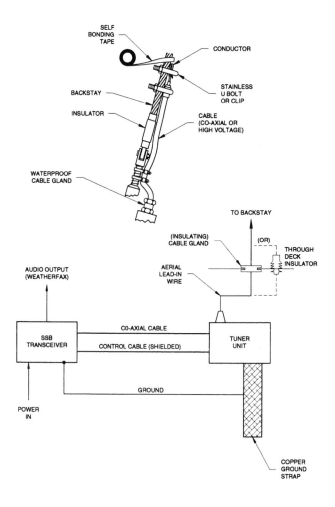

3. **Insulated Backstay.** The insulated backstay is the most common on sailing boats. They do find some use on trawler motor yachts, often in a triatic stay arrangement and they are most efficient in the 2-8 meg range. Losses can occur here as well, as signal radiates into the mast and rigging. It should be at least 11 meters long for an effective aerial. The insulators should be free of chips and have long leakage paths.

SSB Backstay aerial arrangements

TRIATIC ANTENNA

BACKSTAY ANTENNA

BACKSTAY ANTENNA ON A KETCH

About the aerial feed line

The feed line to the aerial is very important as resistance degrades the transmission signals.

1. **Feed line Cables.** Thin conductors and bad joints result in conductor heating and losses. Ideally the cable should not run close to metal decks or hull.

2. **Insulation Quality.** Insulation losses also occur through conductors and deck feed insulators. Use cables with good insulation values; ideally, a silicon insulated high voltage cable should be used.

3. **Deck Transits.** Poorly insulated leads close to metal decks and hull can cause arcing or induction losses. External cables can also leak when the insulation cracks due to the UV effects of the sun. The best system in steel vessels is the use of through deck insulators. These offer long leakage paths and therefore less signal loss. They must be kept clean.

4. **Backstay Connections.** It is imperative that the feed line to aerial connection be made properly. If it is not you will get signal loss that affects transmission and reception.

What types of HF radio ground?

HF radio problems of transmission and reception are often caused by inadequate grounding systems. Remember that the ground plane is an integral part of the aerial system. If it is poor you may not be able to tune properly to required frequencies.

1. **Ground Shoes.** Ground shoes are the most effective method of providing an RF ground on fiberglass and wooden boats. They provide half of the required aerial length and are an integral part of the radiating system. Essentially the porous metal casting provides a large surface area.

2. **Internal Copper Mesh.** Glass and timber vessels may avoid the installation of ground plates by glassing in a large sheet of copper mesh.

3. **Copper Straps.** The interconnecting copper strap from tuner unit to ground plane is essential. It must be a strap, not cable and the surface area is the critical factor. To be effective, a low resistance is required and is the cause of many performance drops and interference. The ground strap should be 2" wide at least. The copper strap should be installed clear of bilge areas.

What HF radio maintenance is required?

There are regular maintenance tasks that will ensure good radio performance.

1. **Aerial Connections.** The lead wire aerial connections should be regularly checked for deterioration. If exposed the wire may degrade, and introduce resistance into the circuit. Always tape up the connection with self-amalgamating tape.

2. **Insulators.** Always clean the insulators to remove salt deposits that encrust and cause surface leakages. This should include the upper insulator on the wire antennas. A damp rag is the best tool.

3. **Ground Connection.** Check the RF ground connections. Clean and tighten the bolts and connection surfaces. After this apply a light smear of petroleum jelly to prevent deterioration in the bilge area. It is advisable to always check and keep this area clean and dry if in a bilge area, as reaction between the copper strap and metalwork can cause corrosion problems.

About HF radio troubleshooting

There is No Reception

Wrong channel selected; Propagation problems; Aerial lead wire broken; Aerial connection corroded; Tuner unit fault

There is Poor Reception

Propagation problems; Aerial connection corroded; Insulators encrusted with signal leakage; Aerial grounding out

There is No Transmission

Tuner unit fault; Aerial connection corroded; Insulators encrusted with signal leakage; Aerial grounding out; Aerial lead wire broken; Ground connection corroded; Low battery voltage; Transceiver fault

There is Poor Transmission

Propagation problems; Aerial connection corroded; Insulators encrusted with signal leakage; Aerial grounding out; Tuner unit fault; Ground connection corroded.

Standard time frequencies

A useful function is the ability to accurately fix time, plus weather data. It is available from the following stations:

1. **WWV (Fort Collins)**. Times are announced at the 8th and 9th minute past the hour on 2.5, 5, 10, 15 and 20 MHz. Information also given on weather, location and movement of storm centers, wind speeds and propagation data at 8, 9 and 10 minutes past each hour.

2. **WWVH (Kekaha, Hawaii)**. Times are announced at the 48th, 49th and 50th minute past the hour respectively on 2.5, 5, and 10 MHz. Information also given on weather; location and movement of storm centers; wind speeds and propagation data at 48, 49, 50 minutes past each hour.

3. **VNG (Llandilo, Australia)**. Times are announced continuously on 5.000, 8.638, 12.984 MHz; 2200-1000 UTC on 16 MHz. Voice broadcasts on 5 and 16 MHz on 15th, 30th, 45th and 60th minute.

6. AMATEUR (HAM) RADIO

About ham radio on boats

Ham radio is the realm of a worldwide group of radio enthusiasts. Ham operators have been involved in many life saving efforts with sailors, but regrettably ham operators and the system have been badly abused. Ham is always useful with earthquakes and other natural disasters and came into its own during the Asian tsunami. The ham operators are unsung heroes who save many lives. Ham radios are a major communication source in the cruising world. In the US, about 70% of cruisers sail with ham, while for the UK and Australia it is probably around 10%. There are a number of important factors to consider.

1. **Operator Licensing.** The operator is licensed, not the station. There are a number of levels that give either partial or full access to frequencies. Levels require tests in Morse code, radio theory, rules and regulations with respect to operations. Fear of technical matters and theory as well as the Morse test frightens off many would be amateurs. A general class license will be required for access to Maritime Mobile Nets in the 15-, 20- and 40-meter bands.

2. **Penalties.** You must be licensed for the country of operation. Be aware that in some third world countries where communications are controlled, jail and vessel loss can occur if it is used in port and without authorization. In many cases you will not be acknowledged on ham bands unless you are licensed and have a call sign.

What are the advantages of SSB vs. Ham?

This is a perpetual argument with both systems having a place on board your boat. You can purchase a combination ham and SSB e-mail ready unit and these days with GMDSS and the demise of shore stations, ham is a clear winner.

1. **SSB.** Radio sets are generally easier to operate, and with automatic tuning, it is simple to punch in a channel number and talk. Additionally radios have automatic emergency channel selection, and are type-approved for marine communications. Only a restricted license or permit is required. You may operate a SSB radio on amateur frequencies if you have a ham license. One of the disadvantages of SSB on ham frequencies is that synthesizers are programmed in 0.10 kHz steps. Ham communications may be at frequencies outside of that so that SSB sets can be marginally off frequency. Most SSB sets operate on upper side band (USB) while most frequencies below 40 meters are lower side band (LSB).

2. **Ham.** The ham operator must have a license appropriate to the frequency band being worked. Access to GMDSS emergency frequencies is illegal except in emergencies. It is illegal to operate non type-approved radios such as ham radios on marine frequencies. Ham allows the use of casual conversation which marine SSB does not. Ham gives full access to information packed nets, and a worldwide communications network. Ham does not readily allow access to telephone networks, although some stations and net controllers may offer phone patches.

Ham radio nets in operation

A good receiver allows listening to ham nets and valuable information as marine SSB sets cannot access them. The following maritime mobile net times could vary an hour either way depending on summer time changes in respective countries. Frequency 14.314 is monitored virtually 24 hours, and is the de facto maritime mobile international calling frequency.

Atlantic/Caribbean/ Mediterranean Nets

UTC	Frequency	Net Name and Area
0100	3.935	Gulf Coast hurricane net
0230	14.313	Seafarers net
0530	14.303	Sweden net
0645	12.353	Greece net
0530	7.088	Eastern Med net
0700	14.313	German MM net
0700	14.303	International net
0800	14.303	UK net
0900	14.313	Mediterranean net
0900	7.080	Canary Island net (Atlantic)
1000	14.303	German net
1030	3.815	Caribbean WX net
1030	14.265	Barbados cruising net
1130	3.815	Antilles emergency weather net
1130	14.320	South Africa MM net (South Atlantic)
1230	7.240	Caribbean net
1300	7.268	Waterway net (US East Coast/Caribbean)
1300	21.400	Transatlantic net (operates in crossing season)
1400	7.292	Florida Coast net
1600	14.313	US Coast Guard net
1700	7.240	Bejuka net (Central America)
1800	14.303	MM net (Atlantic weather forecast)
2030	14.303	Sweden net
2300	7.190	Admirals net (US West Coast)

Pacific/Asia/Indian Ocean Nets

UTC	Frequency	Net Name and Area
0100	21.407	MM net (Pacific/Indian Ocean)
0200	7.290	Hawaii interisland net (MF)
0220	14.315	John's weather net (Norfolk Is and Pacific)
0230	14.313	Seafarers net (also operates Atlantic)
0300	14.106	Travelers net
0300	14.313	DDD (Doers, Dunners and Dreamers) net
0400	14.318	Arnolds net (Weather Pacific)
0500	21.200	Aus/NZ/Africa net (Indian and Pacific Ocean)
0530	14.314	Pacific MM net (Pacific via relay stations)
0630	14.330	Durban net (Indian Ocean)
0630	14.180	Pitcairn net
0700	14.220	Pacific net
0715	3.820	Bay of Islands net (Sth Pacific/Australia)
0800	14.315	Pacific interisland net
1000	14.320	Dixie's net (Philippines, WX NW Pacific)
1000	14.330	Pacific gunkholers net
1200	14.320	SE Asian net
1430	3.963	Sonrisa net (Baja California)
1545	14.340	Marquesas net
1600	7238.5	California Baja net
1630	21.350	Pitcairn net

1700	14.329	MM Hawaii net
1700	14.115	Pitcairn net
1700	14.329	Skippers net
1700	14.340	California Hawaii net
1730	14.115	Jerry's net
1800	14.282	South Pacific net
1800	7.197	South Pacific Sailing net
1900	21.390	MM's Pacific net
1900	7.285	Shamaru net (Hawaii)
1900	14.329	Bay of Islands net
1900	14.340	Manana net (Mexico)
1900	7.288	Friendly net (Hawaii)
1900	3.990	Northwest MM net (NW Pacific)
2000	12.359	Pacific cruisers net (Herb Hilgenberg)
2000	14.305	Confusion net (Pacific)
2030	7.085	Sydney/New Caledonia net
2100	14.315	Tony's net (Sth Pacific MM's Only)
2100	7.060	Coral Coast net (Airlie Beach,Aus)

7. SHORT-WAVE RADIO

Regular monitoring of the news services can often inform you of changes or other factors that may affect your plans. Frequencies may alter marginally. Contact the broadcasters for free schedules and frequency information.

Voice of America (VOA)

VOA broadcasts worldwide. Times are UTC. In many cases you may be able to tune into broadcasts to other areas for limited periods. Frequencies are subject to variation.

1. **Caribbean Service.** Broadcasts 0000-0100 on 5995, 6130, 7405, 9455, 9775 and 11695; 0100-0130 on 5995, 6130, 7405, 9455, 9775, and 13740 kHz; 0130-0200 5995, 6130 and 9455; 0200-0500 1530 and 1580; 1000-1100 on 6165, 7370 and 9590.

2. **Far East/Pacific Service.** Broadcasts on 1143, 1575, 6160, 7115, 7125, 7215, 9635, 9645, 9760, 9770, 9775, 11705, 11760, 15185, 15290, 17740, 17820 kHz.

3. **Europe/Mediterranean.** 792, 1197, 1260, 1548, 6040, 6160, 9530, 9680, 9700, 9760, 9770, 11805, 11965, 15205, 15255 kHz.

British Broadcasting Corporation (BBC) World Service

The BBC World Service is the news service most commercial mariners tune in to, and it is considered the most accurate world news information provider. Time (Frequencies), and time is UTC.

1. **Caribbean Service.** Antigua 98.1 MHz @ 00.00-2400.

2. **Mediterranean Service.** 0200-0400 (9410); 0400-0600 (1323, 6195, 7115, 9410); 0600-1900 (1323, 9419, 12095, 15070, 17640); 1900-2230 (1323, 6180, 6195, 7325, 9410, 12095, 15070).

3. **North America.** 000-0230 (5975, 6175, 9590); 0230-0330 (5975, 6175, 9895); 0330-0700 (5975, 6175); 0900-1000 (6195); 1200-1300 (5965, 6195, 9590, 15220); 1300-1400 (5965, 6195, 9515, 9740, 11865, 15220); 1400-1600 (6195, 9515, 9740, 11865, 15220, 17840); 1600-1800 (17840); 2200-2400 (5975, 6175, 9590).

4. **Pacific SE Asia Service.** This covers Papua New Guinea, Radio Fiji, Radio Tonga, Solomon Islands, Western Samoa, Radio Tuvalu, Radio Kiribati, Radio Vanuatu and Radio Niue. Times are GMT. World News is broadcast at 0100, 0130, 0300, 0600, 0700, 0800, 0900, 1200, 1400, 1500, 1600, 1700, 1900, and 2100. Newsdesk is broadcast at 0000, 0400, 1000, 1800, and 2200. Newshour is broadcast at 1300 and 2200. Mornings tune to 5975, 9740, 11955 kHz. Daytime tune to 6195, 7145, 15360 kHz. Evenings tune to 6195, 7110, 9740, 11955 kHz.

Radio Australia Pacific service

Radio Australia broadcasts to Asia and the Pacific areas. News is broadcast every hour on the hour. Mornings tune to 9415, 5890, and 5995 kHz. Daytime tune to 1180, 7240, 12080, 15510, 17795, 13755, and 12080 kHz. Evenings tune to 15240, 11880, and 9580 kHz.

8. E-MAIL SERVICES

For many "snail mail" is a thing of the past. For most boaters an INMARSAT terminal is not a viable economic alternative, although GMDSS inspired changes make communications improvements essential. If, like myself, you have a quality SSB radio on board, that valuable piece of equipment is your means to get connected to the world.

About HF E-Mail system components

1. **SSB Radio.** Not all SSB radios are configured for e-mail and may require modification to operate, with the addition of an audio output jack. This should provide a line level output signal of 100mV RMS. Radios such as the ICOM M710 are e-mail ready. Radios must be able to transmit full power signal without damage. However older sets including ICOM M700, SEA 235, SGC SG2000 cannot do this, so they must be operated at reduced output power levels. A good power supply is essential to maintain constant transmission and battery voltages must be up and power supply connections sound. Aerials and ATU grounds must also be good to ensure optimum transmission and reception.

2. **HF/SSB Modem.** Modems are generally part of the service providers' systems. Those using other non-service company systems such as packet radio enthusiasts use what is called a Terminal Node Controller (TNC); the most common modems are those from Kantronics such as the KamPlus and the Kam98. A modem has a power input, data port, and radio port, along with operating software. The recommended SailMail modem is the SCS PTC-II, as it is compact and has lower power consumption and faster speeds. The audio cable to the SSB consists of 4 wires: transmit audio (TxD); receive audio (RxD); push-to-talk (PTT); and the audio signal ground. The audio cable must be shielded with the shield being con-

nected at both ends. Prewired cables are available from Kantronics. Clip-on ferrites must be fitted at both ends to reduce RF interference, and also coax line isolators (ungrounded T-4 model), and these are available from www.radioworks.com.

3. **Notebook/Laptop Computer.** Many boaters are incorporating this as an essential part of the equipment inventory so the addition of an e-mail function further enhances the investment.

4. **Software.** Software is required. AirMail is a proven Windows-based message package that uses the Pactor protocol. You prepare messages using the text editor and attach word processing files with point-and-click simplicity, as well as automating the radio link. You can download the AirMail for use with SailMail from the SailMail website. Also go to www.winlink.org/airmail

Transmission system modes and configurations

Both the principal service providers and alternative systems utilize different methods for handling e-mail traffic. Although similar equipment is used, the systems cannot communicate with each other.

1. **Clover.** These modems are used by PinOak and are made by HAL Communications in the US. They use a four-tone signal and are used in the PinOak PODLink-e service. Currently Globe establishes a link in SITOR (marine telex) and then switches over to Clover mode. PinOak does not use SITOR but establishes links either in Clover or PacTOR 2.

2. **PacTOR 2.** These modems are made by SCS in Germany. They use a two-tone signal and are far more effective and reliable with data transfer in noisy environments. They are a hybrid Packet/Amtor modem. They are becoming the favored modem type for use in most marine HF e-mail systems. The new PinOak PODLink-f service utilizes PacTOR modems. PacTOR is replacing Amtor communications due to improved capabilities and is supported by many Aplink stations.

Who are the E-Mail service providers?

There are several ways to connect to e-mail via HF, SailMail and WinLink. They have become very popular and two other main service providers and pioneers of this service are listed below. Both offer GMDSS level services that require installing satellite systems, such as weather and navigational warnings.

1. **Sailmail.** This system operates using SSB radio. It is a relatively cheap and non-profit group and the best option. Check out www.sailmail.com for complete details on using the system and at www.pentacomstat.com.au. The designated frequencies and SailMail stations are as follows, and in J3E mode with non-automatic tuning you will have to subtract 1.7kHz from the listed frequencies.

WRD719, Palo Alto, California; 2661.4, 5881.4, 7971.4, 13971.0, 18624.0 kHz.

KZN508, Rockhill, South Carolina; 2656.4, 5876.4, 7961.4, 7981.4, 10331.0, 13992.0, 13998.0, 18618.0, 18630.0 kHz.

VZX1, NSW, Australia; 6357.0, 8442.0, 12680.0, 16908.0, 22649.0 kHz.

2. **PinOak Digital.** Stations located worldwide include Galá-
   pagos, Falkland Islands, Cape Town, Cape Verde Islands,
   Grand Banks, West Greenland, Eastern Mediterranean,
   Sri Lanka, Hawaii, Tahiti, Wellington, South China Sea,
   Perth, and others. Coverage varies from 5 hours up to a
   full 24 hours and it is for commercial vessels. PinOak Digital, P.O. Box 360, Gladstone, NJ 07934; Tel 800/746-
   6251; Fax 908/234-9685. Users are charged a
   subscription fee, which allows a specific amount of data
   transfer, and then a charge per kilobit transferred. Over
   4000 worldwide weather forecasts are available, along
   with e-mail services and Internet access.

3. **AMTOR** (Amateur Teletype Over Radio). This is probably the cheapest option that I have seen in wide use. The
   system uses what is termed Amtor Packet Link (Aplink).
   These Aplink stations are ham stations configured for automatic reception, storage, and transmission of Amtor
   messages. Messages are transferred between stations until
   the designated destination station is reached. Addressing
   mail requires the recipient MBO (Electronic Mail Box) details. What I found most attractive with on-board systems
   using this system is the ability to "talk" with other vessels
   on a chat net. Log on to www.airmail2000.com and
   www.shortwave.co.uk for useful information and
   www.win-net.org for ham e-mail shore stations.

4. **Globe Wireless.** They are not interested in pleasure boats.
   However they have stations worldwide in San Francisco,
   New Orleans, Hawaii, Bahrain, Sweden, Newfoundland,
   Australia, and New Zealand. Service offered is called
   GlobeE-mail, along with GPS position reporting tied with
   USCG AMVER system. Users are charged a subscription
   fee, which allows a specific amount of data transfer, and
   then a charge per kilobit transferred. Message reception is
   similar with automatic notification.

Alternative E-Mail systems

The main systems offer a seagoing system, but there are other useful options.

1. **WinLink.** For licensed ham radio operators using a Pactor based system, log on for details at www.winlink.org.

2. **Pocketmail.** I have some friends who regularly message me from their boat in Europe using this system. This system can be used with the GSM cell phone. As Palm backs up via a cradle to the laptop all messages are prepared using a full size keyboard then downloaded to the Palm. Log on to www.pocketmail.com or www.stargate3.co.uk.

3. **SeaMail.** This Australian system is operated through Penta Comsat and services the Pacific. Log on to www.xaxero.com for details and software downloads.

4. **CruiseE-mail.** This is a Florida based service (who has a tie up with SeaMail) to offer Atlantic, Caribbean and Pacific Services.

Internet cyber cafés, e-mail centers and wireless hotspots

Many marinas now offer access to e-mail and Internet. Log on to www.ipass.com for information on how to access your ISP back home when you are away by dialing a local telephone number. In most cases, you can find an Internet or cyber café virtually anywhere, look for areas where students go such as universities or near tourist and backpacker hotspots. Then just log on to Yahoo or Hotmail and create an account. I have switched to Yahoo as they have 20 meg accounts against Hotmail and 2 megs, and the latter I found had unsustainable amounts of junk mail and slower access speeds. Also Yahoo can handle large attachments better. It is worth the $10 per year for the increased services.

Where you are paying high Internet access rates a 256 meg or above memory stick is worth its weight in gold and preparing your e-mails before you log on is worth the effort. Then it's cut and paste in your mails.

Some European and US marinas have wireless hotspots although access charges are a little high, this makes Internet access easy while on board.

About acoustic couplers

This system still requires a notebook computer with an appropriate modem card installed, as well as an acoustic coupler. Subscribing to an Internet provider such as Compuserve gives you appropriate e-mail access. Download or send your mail from a phone on land and then you're back off to the boat. It is an economical alternative if you don't mind taking your PC ashore to a phone booth. Wireless hotspots and economical access in Internet cafés is making this method a little redundant.

9. SATELLITE COMMUNICATIONS

There are a variety of services available. The system selected will depend on many factors including the services required, the coverage area, the initial installation costs, the size of the antennas, and the annual and call costs. Log on to www.heavens-above.com for information on satellite orbits and tracking.

All about INMARSAT

INMARSAT is based on INternational MARitime SATellite Organization. The system comprises four satellites in geostationary orbit, 23.000 miles (36,700km) high, and the satellites remain in the same position relative to the earth. There are 4 ocean regions (Atlantic East, Atlantic West, Pacific and Indian). The ship station or Mobile Earth Station transmits and the satellite relays the signal to a Land Earth Station (LES) for routing via terrestrial communications networks. Log on to www.inmarsat.org for details.

1. **Standard-A SES.** The first system introduced in 1982 and unlike later systems is analog. New standard A systems have dramatically decreased the size of equipment. It supports telephone, telex, fax, e-mail and data transmissions. It also supports GMDSS requirements, but is seen only on large commercial vessels due to the large size of the antenna.

2. **INMARSAT-B.** This was introduced in 1994, and will replace Inmarsat A by 2010. This is a high-speed digital service that provides high quality data, fax, telephone, telex, video transfer and conferencing. Inmarsat B HSD (High Speed Data) is a more recent enhancement and transmits data at 64k bits/sec direct to an ISDN line. This system is now a primary one for many commercial vessel operations and complies with all GMDSS requirements.

3. **INMARSAT-C.** This worldwide service was introduced in 1991. It provides reliable data and is an essential part of GMDSS. It uses small antennas and a notebook computer. It can be programmed to receive Enhanced Group Calls (EGC) and then get SafetyNet broadcasts of Marine Safety Information (MSI). The system uses 4 geostationary satellites in addition to terrestrial services. You can send and receive e-mails using a Windows Outlook format. Costs are based on receipt and sending per character. You have limits on message sizes and no attachments can be sent.

4. **INMARSAT-D+.** This new service provides global 2-way paging services with a limit of 128 characters per message. It is also utilized as a vessel-tracking device.

5. **INMARSAT-M and Mini-M.** This service was introduced in 1992 and enhanced in 1997 with the new Series 3 satellites placed into orbit. The system offers secure digital voice, fax, data (2.4Kbps) and e-mail. Antennas are small at around 10 inches (25cm), gyro-stabilized, and have become a very popular system on many powerboats. SIM card pre-paid call billing options are offered. The system is global within the Inmarsat Mini-M satellite area and utilize "spot beam" technology. The signal is beamed to specific areas, such as main landmasses and coastlines, with coverage being typically up to 200 nm offshore.

6. **INMARSAT-P.** Currently under development, it will offer global handheld system with voice, paging, fax, and data services.

Other satellite systems

1. **Globalstar.** Consists of 48 low earth orbit (LEO) satellites, which are dual mode satellite and GSM 900 systems. Services include high quality voice, short messaging services (SMS) and roaming. Also dial-up fax and data services are available at 9.6 kbps (bits per second). Coverage is limited to coastal areas to around 200 nm offshore, which will suit many people. This is not a GMDSS system. The domes are very small.

2. **Motient/AMSC.** This system has a single geostationary orbit satellite and has a footprint covering North America and Hawaii up to 200 nm offshore. It offers voice, fax and e-mail services.

3. **ORBCOMM.** Consists of 26 LEO satellites. The service is digital data only and will offer paging, e-mail, etc. It is not a GMDSS system.

4. **Iridium.** It consists of 66 satellites in low earth orbit (LEO) and offers global voice, data, e-mail, fax, and paging services. The system is now operated by Boeing for the US Department of Defense. This is not a GMDSS system. Iridium offer enhanced mobile satellite services (EMSS) that give low rate data and voice; the data rate is at 2.6 kbps. This is not good for web surfing or large e-mail attachments. You can use Iridium Gateway that will provide 9.6 kbps using data compression and this requires software. The iridium unit is also a small and compact handheld unit. You can install a fixed system with better antenna and therefore better performance. The system is not GMDSS, but you can on the Skanti ScanSet system have a dedicated alter button that auto dials the nearest Iridium Gateway which routes the call to any RCC (Rescue Coordinating Center).

5. **VSAT.** (Very Small Aperture Terminals) They use Ku-band geostationary satellites such as Intelsat and Eutelsat with data rates of 1.5 mbps. They do not have global coverage with any one company, and require more precise antenna tracking.

6. **Thuraya.** This system has a single geostationary orbit satellite and a footprint similar to the Inmarsat Indian Ocean unit. The handsets offer voice and data services. The unit is a compact handheld phone that is ideal for those cruising the Mediterranean, Middle East and Europe.

7. **Emsat.** This new system also has a single geostationary orbit satellite and the major satellite communications company Eutelsat operates it. Coverage is limited to the Mediterranean and Northern Europe.

8. **ICO.** Intermediate Circular Orbit, uses 10 Medium Earth Orbit (MEO) satellites, offering voice, fax and data.

9. **Teledesic.** This system will use 288 low earth orbit satellites. The system is a broadband Internet-in-the-Sky. Ka-Band radio waves uplink 28.6-29.12 GHz at 2 Mbps and 18.8-19.3GHz downlink at 64 Mbps. The system will offer high-speed Internet access, interactive multimedia and high quality voice communications.

About satellite system installation

The following is for a Nera Saturn Bm system, and the criteria apply to all systems. Systems consist of the radome, which encloses a stabilized antenna dish, a pedestal control unit (PCU) and the RF Unit. Follow installation instructions in the user's manual precisely as warranty may be voided if installation is incorrect.

1. **Radome Installation.** The radome must be located as far as practicable from any HF and VHF antenna, and preferably at a minimum of 5 meters from all other communications and navigation receiver antennas. Avoid locating near an exhaust funnel, as soot will gradually degrade performance. Do not mount the unit in any location subject to vibration. Safe compass distance is a minimum of 1 meter. Systems also require radiation precautions, and should be 5 meters from any accessible area, and 2 meters above to avoid excessive microwave radiation. If at 2 meters vertical clearance the 5-meter rule is not required. The radome should be outside the beamwidth of radar antennas, typically 10 degrees each side of the central plane. The radome should be properly aligned parallel with the ship's axis. As beamwidth is 10°, a clear line of sight is required from 5° elevation and above. Obstructions will create blind spots, and disrupt communications. Obstructions less than 15cm are acceptable within 3m of the antenna, but marginal signal strengths are vulnerable to them. The azimuth and elevation angles must be considered at all times. Normal cable installation rules apply, and must be observed to prevent mechanical damage. The antenna unit uses double-screened 50-ohm coaxial cable. This is usually RG223/U and RG214, with maximum lengths of 13m and 25m respectively to achieve 10dB/0.6 ohm maximum losses and attenuation. All cables must be shielded, and the shield grounded. Peripheral equipment must also be grounded. Co-axial connectors must be put on correctly, and this is a frequent cause of problems.

2. **Operation.** The antenna can be directed accurately at the
nominated satellite to optimize signal transmission and reception. In normal operation, the dish auto-tracks the
satellite. To do this the dish must be aligned correctly. The
nominated satellite is based on ship position, and then selection of a relevant satellite with area coverage. The vessel heading is required to give correct azimuth heading,
and gyro or fluxgate input provides this. The azimuth
angle is the angle from North and horizontal satellite direction. The elevation angle is the satellite height above
the horizon in relation to the vessel. At power-up, the system must locate a satellite and synchronize with it. This is
either by automatic or manual initiated hemispheric scan
for selected or ocean region satellite. The dish does a
search pattern until the satellite signal is located in the relevant ocean region. Systems also carry out a self-test at
initialization. Systems default to last settings on gyro, azimuth and elevation, and if the vessel position is lost, this
data is required. Under GMDSS the default LES and Distress Alarm address must be configured. During operation, displays on handsets show signal quality, and signal
strength Signal/Noise Ration (S/N). Bit error rates (BER)
decrease with increased signal quality.

10. WEATHERFAX

About weatherfax receivers

Weather facsimile gives access to many stations that transmit weather charts and the charts are a lot easier to interpret than foreign language voice forecasts. Weatherfax services in the US, UK and Europe are under threat amid cost cutting in the wake of GMDSS implementation. Transmitted data is varied and includes ocean current positions, sea temperatures charts, and current weather charts every six hours, forecasts up to five days in advance, sea state and swell forecasts and ionosphere propagation forecasts.

ICS Weatherfax

Weatherfax facsimile signal components

A facsimile transmission consists of a number of distinct components:

1. **Continuous Carrier.** This single tone is emitted before the start of any broadcast. It allows the receiver to be tuned to maximum signal strength prior to data reception.

2. **Start Tone.** This is also called the index of co-operation (IOC) select tone. It enables receivers to recognize the start of a transmission and to select the appropriate IOC drum speed.

3. **Phasing Tone.** This tone synchronizes the edge of the transmitted image.

4. **Scale Tone.** Some systems enable the tone variations within the broadcast to be selected or varied.

5. **Body of Transmission.** This characteristic rhythmic "crunching" tone is the facsimile data being decoded into an image.

6. **Stop Tone.** The stop tone is similar to a start tone and indicates the end of the transmission.

7. **Close Carrier.** This tone follows conclusion of the transmission.

About weatherfax decoders

To obtain weatherfax data it is necessary to obtain signals via a SSB or short-wave radio, and decode them for display on a laptop computer, or printer as required. The basic function of a decoder is to convert transmitted audio signals into data. The audio signal is taken from the audio jack if fitted or a terminal on the rear of the SSB set.

Weatherfax printers

An ink-jet printer can be utilized from the laptop computer. As a plain paper printer, it is significantly cheaper to operate than thermal paper roll types. One of the factors to consider is both the ease of printing and the size and quality required. Make sure you carry enough spare paper and ink cartridges for your trip, as these items are often hard to procure.

Discrete weatherfax systems

On larger vessels an integrated decoder and printer system is often used. The most common are Furuno and the smaller ICS Fax-2. A paper roll lasts a considerable period. The unit also has a number of useful features. An additional aerial can be added for full Navtex reception, marine page can be used, and the reception of RTTY and FEC signals is possible. Like most weatherfax units, programming of specific reception times is possible which takes the worry out of looking up and catching broadcast times.

Weatherfax power consumption

The power consumption rate is relatively low, although it should take into account SSB consumption as well if a unit is left on permanently to capture programmed transmissions. If power consumption is an issue, you will have to power up before the required broadcast and after receiving shutdown again. The combination of decoder and SSB over 24 hours can be at least 25-30 amp hours, which is considerable. Typical power rates are as follows:

1. **Standby Listening Mode.** The ICS Fax-2 unit has a drain of only 2.5 watts. The SEA SSB unit power consumption is 2 amps, while the 322 model is only 1.0 amp.

2. **Print Mode.** The power drain increases to approximately 4 amps when printing. The SSB drain remains the same unless the audio is turned up and it can be around another 0.5A.

Weather Facsimile Frequencies

Station	Frequency
Kodiak (USA) NOJ	2054, 4298, 8459, 12412.5
Point Reyes (USA) NMC	4346, 8682, 12786, 17151.2, 22527
Honolulu (Hawaii) KVM70	9982.5, 11090, 16135
Boston (USA) NMF	4235, 6340.5, 9110, 12750
New Orleans (USA) NMG	4317.9, 8503.9, 12789.9, 17146.4
Halifax (Canada) CFH	122.5, 4271, 6496.4, 10536, 13510
Northwood (UK) GYA	2618, 4610, 8040, 11086.5
Northwood (Gulf) GYA	3289.5, 6834, 14436, 18261
Charleville (Aust) VMC	2628, 5100, 11030, 13920, 20469
Wiluna (Australia) VMW	5755, 7535, 10555, 15615, 18060
Wellington (N. Zealand)	3247.4, 5807, 9459, 13550.5, 16340.1
Cape Naval (Sth Africa)	4014, 7508, 13538, 18238
Nairobi (Kenya)	9044.9, 17447.5
Offenbach (Germany)	3855, 7880, 13882.5
Skamlebaek (Denmark)	5850, 9360, 13855, 17510

Rome (Italy)	4777.5, 8146.6, 13597.4
Athens (Greece) SVJ4	4481, 8105
Moscow (Russia)	3830, 5008, 6987, 7695, 10980, 12961, 11617
Tashkent (Uzbekistan)	3690, 4365, 5890, 7570, 9340, 14982.5
JJC (Japan/Singapore)	4316, 8467.5, 12745.5, 16971, 17069.6, 22542
Tokyo (Japan) JMH	3622.5. 7305, 13597
Beijing (China) BAF	5526.9, 8121.9, 10116.9, 14366.9, 16025.9, 18236.9
Shanghai (China) BDF	3241, 5100, 7420, 11420, 18940
Seoul (Korea) HLL2	5385, 5857.5, 7433.5, 9165, 13570
Taipei (Taiwan)	4616, 5250, 8140, 13900, 18560
Bangkok (Thailand)	7396.8, 17520
New Delhi (India)	7404.9, 14842
Rio de Janeiro (Brasil)	12665, 16978
Valparaiso (Chile) CBV	4228, 8677, 19146.4

NOTE: In upper side band mode, adjust frequency 1900Hz (1.9kHz) lower, if in lower side band mode adjust 1900Hz (1.9kHz) higher.

INDEX